CALIFORNIA NATURAL HISTORY GUIDES

FIELD GUIDE TO OWLS OF CALIFORNIA AND THE WEST

California Natural History Guides

Phyllis M. Faber and Bruce M. Pavlik, General Editors

Field Guide to OWLS of California and the West

Hans Peeters

Illustrations and photos by Hans Peeters

UNIVERSITY OF CALIFORNIA PRESS
Berkeley   Los Angeles   London

For my wife and my sons.

University of California Press, one of the most distinguished university presses in the United States, enriches lives around the world by advancing scholarship in the humanities, social sciences, and natural sciences. Its activities are supported by the UC Press Foundation and by philanthropic contributions from individuals and institutions. For more information, visit www.ucpress.edu.

California Natural History Guide Series No. 93

University of California Press
Berkeley and Los Angeles, California

University of California Press, Ltd.
London, England

Library of Congress Cataloging-in-Publication Data

Peeters, Hans J., 1937–.
Field guide to owls of California and the West / Hans Peeters.
p. cm. — (California natural history guides ; 93)
Includes bibliographical references and index.
ISBN 978-0-520-24741-3 (cloth : alk. paper) — ISBN 978-0-520-25280-6 (pbk. : alk. paper)
1. Owls—California. 2. Owls—West (U.S.) I. Title.
QL696.S8P44 2007
598.9'7—dc22
2007005383

Manufactured in China
10 09 08 07
10 9 8 7 6 5 4 3 2 1

The paper used in this publication meets the minimum requirements of ANSI/NISO Z39.48–1992 (R 1997) (*Permanence of Paper*).

Cover: Northern Saw-whet Owl *(Aegolius acadicus)*. Painting by Hans Peeters.

The publisher gratefully acknowledges the generous contributions to this book provided by

the Gordon and Betty Moore Fund
in Environmental Studies
and
the General Endowment Fund of the
University of California Press Foundation.

CONTENTS

PREFACE

Soft, mellow hoots drift from a stand of trees down the block on this bright April morning. "Is that an owl?" your child asks. Do you know the answer?

Because of the secretive and chiefly nocturnal habits of owls, many people, although they can recognize one, have never seen an owl in the wild. Most urban humans are unaware that owls may live in their midst, going about their business in the dark after we have turned on the lights that so effectively shut off contact with the natural world.

No, the calling bird was not an owl. It was a male Mourning Dove *(Zenaida macroura)* with romance on his mind. Owls rarely call by day, particularly when it is sunny. But had you listened after nightfall, you might have heard the trill of a Western Screech-Owl *(Megascops kennicottii)* coming from those very same trees, or the disembodied rasp of a Barn Owl *(Tyto alba)* somewhere in the star-laced sky.

The purpose of this book is to acquaint the reader with a group of birds that are often regarded as enigmas even by those who do see them occasionally. Besides providing information helpful in identifying the various species, the book includes suggestions as to how to find these private creatures, details about their feathers and body structure (with special emphasis on their extraordinary senses of sight and hearing), and much more.

My intent is to present a selection of facts (and anecdotes) about owls that will heighten the understanding of these singular birds and enhance the pleasure of learning about them and watching them. Although it obviously is not the purpose of this guide to serve as an encyclopedic scientific reference, this book is also aimed at professional biologists, who will find useful information they may not have previously known. In discussions of topics not deemed common knowledge, free use has been made of the available scientific literature, complete with selectively chosen citations, so that the interested reader can pursue a given topic to learn more detail. Here and there I have included my own observations and occasionally bold speculations without citation. Some species of owls are well studied, but there are large gaps in our knowledge about others. For example, little is yet known about some of the details of breeding in pygmy-owls *(Glaucidium)*, which are most commonly seen during the nonbreeding season. Perhaps some readers will be inspired to go in search of these elusive gnomes—they are more numerous than is generally believed.

My original plan was to write a volume about owls as a companion to my field guide on diurnal birds of prey, *Raptors of California* (Peeters and Peeters 2005). Once my editors realized, however, that the number of owl species found in California was only a few shy of the total number of species found in North America, they suggested I venture beyond California's borders, and so the text expanded, funguslike, into the other western states. Although there is clearly a California focus—not surprisingly, because I live here and know the local wildlife best—the new scope provided the opportunity to write about species that do not occur in my home state but that I have had the pleasure of meeting elsewhere.

Of the 19 species of owl in North America, 15 occur (or may occur) in California. Missing from my state but found in some other western states are the Northern Hawk Owl *(Surnia ulula)*, which has been recorded as far south as Oregon; the Eastern Screech-Owl *(Megascops asio)*, an eastern species that makes it to Montana, Wyoming, and Colorado; and the largely Latin American Whiskered Screech-Owl *(M. trichopsis)* and Ferruginous Pygmy-Owl *(Glaucidium brasilianum)*, whose ranges include parts of the Southwest.

Although the Snowy Owl *(Bubo scandiacus)* is only a visitor to some western states and does not breed south of the Artic, it is certainly part of the Western avifauna, and a spectacular one at that; news of a sighting of (or, much more seldom, an invasion by) this owl spreads quickly through the bird-watching community and causes pilgrimages to viewing sites. The Elf Owl *(Micrathene whitneyi)* is extremely rare in California and may in fact no longer reside here, but it is too soon to pronounce it extirpated: there are persistent rumors of sightings from the lower Colorado

River, and the species certainly is not uncommon in parts of Arizona, New Mexico, and Texas.

The suggestion of the Boreal Owl *(Aegolius funereus)* as a California species is more controversial and is based on a single observer's hearing its call in midwinter in our state. An inhabitant of high mountains, this owl is most vocal at a time when its habitat is covered in deep snow and owl watchers are toasting their toes by the fire at lower elevations. Many believe that Boreal Owls will eventually be discovered breeding in California; they are resident in central Oregon and are found as far south as northern New Mexico and occur in other western states.

Owls are memorable creatures, a happy fact for someone who wants to write a book on these wondrous birds and who is blessed with a wide circle of friends, many of them biologists, all of them people who go through the land with open eyes and minds.

My most profound thanks go to my best friend and wife, Pam, my constant companion walking the fields and forests of this world, support system extraordinaire, sublime cook, published fellow scribe, editor, typist, and general factotum par excellence who contributed immeasurably to the production of this guide. Too bad she cannot paint birds.

Near-lifelong friends have discussed owls and other raptors with me ad infinitum. My thanks to Sterling Bunnell, Steve Herman, Ed Hobbs, Grainger Hunt, E. W. Jameson Jr., and Bob Risebrough for exchanging ideas and observations over the years, and for their enduring friendship. Bruce Mahall additionally has not only spent a great many pleasant days and nights in the field with me, at times looking for owls, but has also helped to finish many a bottle of wine, often improved by the place of consumption.

I am very indebted to people who have provided me access to live owls or frozen specimens or directed me to roadkills (of variable freshness). These include Nancy Anderson, Rose Britton, Seth Bunnell, Julie Burkhart, Rob and Julie Cyr, Allen Fish (who also helped in many other ways), Pat and Phil Gordon, Susan Heckly, Frank Marino, and Chris Peeters. Carla Cicero and James Patton (who was always available and willing to answer questions) facilitated examining material in the University of California at Berkeley's Museum of Vertebrate Zoology collection.

My special thanks go to Pete Bloom and Brian Woodbridge, who not only carefully read the manuscript, making numerous helpful suggestions, but who also supplied information not yet available in the literature. I also thank Doug Bell for his reading of the first draft.

Irv Tiessen graciously provided two owl boxes for my yard (a Western Screech-Owl promptly moved into one of them, favoring it over my own antiquated model) and was always happy to help in a variety of ways. Steve Simmons, astute observer of owls, generously allowed me the use of his slides, Barn Owl data, and nest box design.

Many other fans of owls, both amateur and professional, have greatly contributed to this book by providing help or information on topics too numerous to list, running the gamut from the use of owls in Italy to the fate of a California Towhee *(Pipilo crissalis)* on the high seas. They include Bud Anderson, Jack Barclay, Joelle Buffa, David Camilleri, Manuel Carrasco, the late Howard Cogswell, Daniele Colombo, D. J. Correos, Kate Davis, Penelope Delevoryas, Joe DiDonato, Leon Elam, Margaret Emery, Sandy Ferreira, Reiko Fujii, Kimball Garrett, Fred Gehlbach, Gordon Gould, Andrea Henke, Buzz Hull, Terry Hunt, Lloyd Kiff, Tim Koopmann, Hans Kruger, Lynn Kruger, Colleen Lenihan, John Loft, Sarah Lynch, Susan Magrino, Michelle Manhal, Jeff Maurer, Geoff Monk, Richard Montgomery, Joe Naras, Claudio Peccati, Hans-Josef Peeters, Julian Peeters, Bob Power, Dave Quady, Pat Redig, Keith Richman, Patricio Robles Gil, Ron Schlorff, John Schmitt, Debi Shearwater, Andy Smoker, Sam Smoker, Scott Stender, Chris Stermer, Diane Tiessen, George Trabert, Marilyn Trabert, Rodney Tripp, Jim Turner, Brian Walton, and Wendy Winstead.

Doris Kretschmer and Jenny Wapner at the University of California Press championed the writing of this book, and the superb editorial skills of Scott Norton and Kate Hoffman and their great many helpful suggestions are deeply appreciated.

Finally, I would like to thank owl researchers in general, for spending hours in the cold at times of the night when sane humans are home in bed.

Map 1 (facing page). Major geographic and vegetational features of the western United States.

Mixed grasslands
Spokane
Palouse grasslands
CASCADE RANGE
BITTERROOT RANGE
ROCKY MOUNTAINS
Billings
Portland
Boise
COAST RANGES
Coniferous forests
Wyoming Basin shrub steppe
Grasslands
Snake shrub steppe
Klamath Siskiyou forests
Salt Lake City
Denver
Great Basin sagebrush steppe
Uinta and Wasatch montane forests
SIERRA NEVADA
Central Valley croplands and grasslands
Sacramento
Short grass prairie
COAST RANGES
Mojave Desert
Las Vegas
Arizona montane forests
Chaparral
Santa Fe
SACRAMENTO MOUNTAINS
Sonoran Desert
Channel Islands
Tucson
PACIFIC OCEAN
N
0
200
400 miles
0
400 kilometers

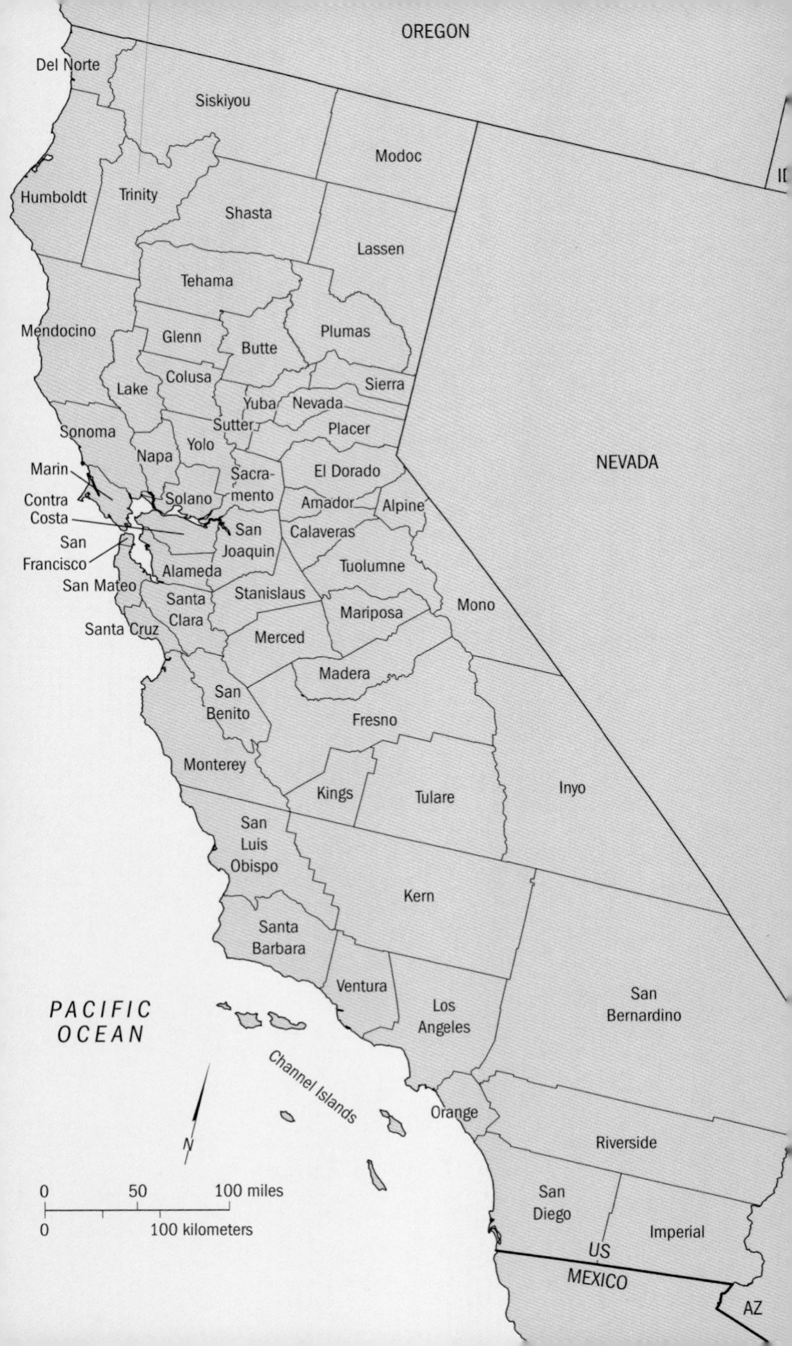
OREGON
Del Norte
Siskiyou
Modoc
ID
Humboldt
Trinity
Shasta
Lassen
Tehama
Mendocino
Glenn
Butte
Plumas
Colusa
Lake
Sierra
Yuba
Nevada
Sutter
Placer
Sonoma
Yolo
Napa
Marin
Sacra-
mento
El Dorado
NEVADA
Contra
Costa
Solano
Amador
Alpine
San
Joaquin
Calaveras
San
Francisco
Tuolumne
Alameda
San Mateo
Stanislaus
Santa
Clara
Mariposa
Mono
Santa Cruz
Merced
Madera
San
Benito
Fresno
Monterey
Inyo
Kings
Tulare
San
Luis
Obispo
Kern
Santa
Barbara
Ventura
San
Bernardino
Los
Angeles
PACIFIC
OCEAN
Channel Islands
Orange
N
Riverside
0
50
100 miles
0
100 kilometers
San
Diego
Imperial
US
MEXICO
AZ

Map 2 (facing page). California counties. When counties are named in this book, they are in California.

INTRODUCTION

MANY BIRDS HAVE ADAPTATIONS so astonishing as to test our credulity. The Whooping Crane *(Grus americana)* has a nearly five-foot-long windpipe, partly coiled beside its breastbone, to produce stentorian, distance-spanning sounds that remind us of trombones, and the kiwi *(Apteryx)*, a rotund flightless bird of New Zealand, smells its quarry through the nostrils at the tip of its long beak while truffling for worms.

But owls stand out with their ingeniously modified plumage, their telescope eyes, their sound funnels, their very skulls sometimes twisted asymmetrical by evolution to facilitate three-dimensional hearing—all this, plus finely honed weaponry. To obtain food in the dark presents no difficulty for a deer, but when that food is not only highly mobile but also has an array of sharp senses to detect a predator, the nocturnal raptor's body requires some extraordinary modifications.

Owls are surely among the most readily recognized groups of birds in the world. Being so distinctive, it comes as no surprise that in bird classification, owls are assigned their own order, the Strigiformes, the owl-shaped birds, separate and distinct from, for example, the Falconiformes (hawk-shaped birds).

What Is an Owl?

Linnaeus, father of animal classification, first placed owls with the hawks, eagles, and falcons because of their hooked beaks, talon-bearing toes, and predatory lifestyle. Most ornithologists today are in agreement that owls are actually most closely related to the Caprimulgiformes (the nightjars, nighthawks, and their allies). With these, owls have in common the often large eyes, the soft, lax plumage with intricate patterning, the rather austere colors, and anatomical and behavioral similarities (some really odd, such as grooming combs and rocking side to side when alarmed), as well as molecular (DNA) affinities. The traits owls share with hawks are merely an excellent example of convergent evolution, driven here by similar food habits. But just when it seems safe to declare owls nothing more than nightjars gone bad, additional molecular studies on the mitochondrial DNA of owls suggest that these birds in fact are related to neither caprimulgids nor hawks (Wink and Heidrich 1999).

Fig. 1. The Neotropical Northern Potoo *(Nyctibius jamaicensis)* (left) and the Common Poorwill (represented below by chicks) are members of the order Caprimulgiformes, a group of birds that shows many affinities to owls on molecular, anatomical, and behavioral levels.

Whereas the caprimulgiform birds are chiefly insectivorous, owls would appear to have evolved in response to an incredibly energy-rich bonanza of food available at night: the rodents. Many modern owls also are insect-catchers and eat other prey (a few even live on fish), and some fly by day, but many are clearly designed to catch rats and mice in the dark. Worldwide, there are over 200 species of owls, separated into two families, the Tytonidae and the Strigidae, the latter comprising the great majority.

Ten thousand years ago, there lived in Cuba a huge kind of owl that stood taller than three feet; it is presumed to have fed on rodents the size of pigs. The fossil bones of this bird indicate that it was a kind of barn owl, a member of the family Tytonidae, whose history can be traced back to the Miocene, 26 million years ago, far before the appearance of hominids. Modern tytonids, substantially more modest in size, are distinctly warm-temperate to tropical in distribution, suggesting that the family originated in warmer regions. As opposed to North America, where the Barn Owl *(Tyto alba)* is the sole representative of the family, Australia, for example, with an overall balmier climate, is home to five species of barn owls, including one, the Masked Owl *(T. novaehollandiae)*, that looks like a Barn Owl on steroids and catches rabbits with its powerful feet. It is of course still vastly smaller than the extinct Cuban giant. The various members of the family are sufficiently similar to the Barn Owl that they are readily recognized as relatives, with the exception perhaps of the bay owls (for one, see the photo of the author), two tropical species found in jungles and fittingly endowed with exotic-looking faces.

The various kinds of barn owls worldwide share distinctive anatomical traits such as a heart-shaped facial disk, long legs, and, in most, a comblike middle claw (pecten) used for grooming. When molting, they replace tail feathers from the tail's center out

Fig. 2. The pecten (comb) on the Barn Owl's middle talon is used for grooming feathers; it is also present in the Common Poorwill.

toward the edges, the opposite of the tail molt of strigid owls (Gilliard 1958), which also have more circular facial disks (usually resembling in outline the cut surface of a halved apple), mostly shorter legs, and no pecten. In addition, no tytonid owl has ear tufts, a feature of many strigids. In captivity, fertile hybrid eggs have been produced by members of the two families, but no young hatched (Flieg 1971).

Owls of the family Strigidae are not only represented by vastly more species but occur in a far greater range of sizes and appearance. From the pygmy-owl *(Glaucidium)*, scarcely larger than a sparrow, to the massive eagle-owl *(Bubo)*, these birds occupy every major terrestrial habitat on Earth and are found on every continent except Antarctica. Secretive as they are, it comes as no surprise that new species are still being discovered, and DNA and vocalization studies may reveal that a known kind of owl actually hides a second species within its ranks, virtually identical in appearance.

Owls in California and the West

All the owls native to North America north of Mexico can be found in the West, and California alone is home to more than a dozen different kinds, placing it at or near the top in species abundance when compared to other U.S. states, although a few types are very rare here.

No California county is without two or three species, and most have substantially more; some owls are in fact common and live comfortably in the presence of humans. In the oak-laurel and oak–gray pine woodlands of the Coast Ranges and places where these habitats have been integrated into suburban developments, as many as six species can be heard calling and singing on a night in late fall, winter, or early spring: the Great Horned *(Bubo virginianus)*, Long-eared *(Asio otus)*, Western Screech- *(Megascops kennicottii)*, Barn, Northern Pygmy- *(Glaucidium gnoma)*, and Northern Saw-whet *(Aegolius acadicus)* owls all may announce themselves during a walk along an otherwise silent road after nightfall.

Such a stroll can develop into an exciting outing. The aspiring owl watcher soon discovers that owls produce a bewildering array

Fig. 3. Female Burrowing Owl and some of her prefledged young, one lying down to rest, another turning its head sideways, behavior the function of which is not clear.

Fig. 4. Tule Lake and the Klamath Basin in northern California are famous for their wintering concentrations of waterfowl, Bald Eagles *(Haliaeetus leucocephalus)*, and other raptors. It is an excellent place to see Short-eared Owls *(Asio flammeus)* as well as some other owl species.

Fig. 5. The towering redwoods of California's north coast are famously home to Northern Spotted Owls *(Strix occidentalis caurina)* and also to several other species.

of vocalizations, including some that seem to have sprung from the imaginations of the brothers Grimm, turning the dark wood into an abode of trolls and gremlins. Even an experienced owler can be unnerved temporarily by what sounds to be the scream of a murder victim in extremis. And there are other nocturnal surprises: the sharp-antlered Black-tailed Deer buck *(Odocoileus hemionus)*, confused by your flashlight's beam, approaching within touching distance; the Bobcat *(Felis rufus)* that slips into the roadside bushes with a fire-eyed backward glance; the Common Poorwill *(Phalaenoptilus nuttallii)* whirling up on moth wings from the road; and of course, always, the rustling in the brush that is entirely too loud to be made by just a rabbit.

Conversely, for the more leisure-oriented owl watcher, there is the unalloyed pleasure of sitting in the comfort of your car on a bright morning in late spring and watching the endlessly amusing carryings on of a family of Burrowing Owls *(Athene cunicularia)* at the mouth of their burrow; and coming across a roosting owl during a brilliant summer's day walk in the park is always a happy event. With experience, you can find your favorite birds in just about any habitat, from the Anza-Borrego Desert to the oak woodlands of Pinnacles National Monument and the coniferous fog-shrouded gloom of the northwest coast, and your encounters need not involve strenuous exercise or after-dinner forays, though these tend to be the most fruitful.

Fig. 6. This modest strip of oak riparian woodland along an ephemeral stream amidst grassland in central California measures less than 1 km (less than .5 mi) in length. In August 2005, it held eight Barn Owls, one pair of Great Horned Owls, and at least one pair of Western Screech-Owls. Circled UFO is a flying Barn Owl.

Most owls are closely associated with trees of some sort, and their bodies reflect this. And while, like hawks, they are birds of prey, they show much less diversity in hunting methods and in body and wing shapes than do the diurnal raptors. Because the foraging habitats and foraging styles are so similar in most owls, they are remarkably alike in build, though some have longer wings than others or longer legs. Notwithstanding the differences in markings between the various western species, they all share much the same similarly distributed colors, and only three defy this convention (the Barn Owl, the Great Gray Owl *[Strix nebulosa]*, and Snowy Owl *[Bubo scandiacus]*); one, the Northern Hawk Owl *(Surnia ulula)*, abandons the traditional shape and looks more like a hawk.

AN OWL'S BODY

THE OVERALL APPEARANCE OF AN OWL is so familiar that probably most people on the planet can recognize one, whether it be the 40 g (1.4 oz) Elf Owl *(Micrathene whitneyi)* or the Snowy Owl *(Bubo scandiacus)*, which may weigh as much as 2,951 g (over 6.5 lb). Even city dwellers can tell you that an owl is a rather plump, big-headed bird whose enormous, forward-directed eyes are surrounded by what appears to be a small satellite dish antenna, but few people are aware of the marvelous complexity of those unique owlish features.

Owls are sometimes seen after dark as seemingly disembodied wraiths that ghost through headlight beams as a car hurtles down the night-black I-5 or some other highway. Close-up or in hand, however, most owls are not at all ghostlike but rather are creatures of substance that more aptly resemble a miniature samurai. Usually short and stocky, their compact bodies are dressed in finely detailed attire, and they are heavily armed; several species bear hornlike feather tufts on bulbous heads that recall a warrior's helmet. In action, the great eyes blaze as the raptor scans the darkness to locate prey: like the vassal swordsman, an owl gives no quarter.

This rather startling appearance is the outcome of exquisite evolutionary fine-tuning of a bird's body needed to secure nimble, alert prey in the dark of the night. The most important modifications are the silencing of noisy feathers while in flight, the development of eyes capable of functioning under very low light conditions, and the ability to locate objects in complete darkness by sound alone.

An Owl's Head

The enormous-appearing head of an owl looks like it should be full of brains. It is not. The skull is conspicuously broad, providing the large eyes with room to face forward and at the same time widely separating the ears (not to be confused with the ear tufts projecting above the head), which facilitates differential hearing (see "Hearing" section).

A thick but light layer of sound-muffling feathers makes an owl's head appear even larger, particularly in species that have a well-developed facial disk, which, as a structure associated with hearing, is much less prominent in the few owls that hunt mainly by sight. In the Barn Owl *(Tyto alba)*, which has relatively small eyes, the skull is narrower and relatively longer than that of other owls; but this species is nevertheless a superb auditory hunter and has a prominent and large facial disk. Interestingly, a Barn Owl's skull is very reminiscent of a dolphin's parabolic skull.

Fig. 7. Head of a Barn Owl chick, about two weeks old. This species has a longer and narrower head than other owls.

Besides varying in size, the facial disk also differs between species in shape and in development. A well-developed disk consists of a variety of highly modified small feathers, some with only skeletal barbs (the fine branches of a feather that form the flat surfaces); they appear designed to permit sound passage toward the stiff, curved, scalelike feathers that arise from skin flaps behind the ears and that form the conspicuous border of the disk, the so-called facial ruff, which directs sound into the owl's actual ears. Northern Pygmy-Owls *(Glaucidium gnoma)*, which forage chiefly by sight rather than by hearing, have much smaller and not very elaborate facial disks that are more elongate and bring to mind stylish eyeglass frames. They also have relatively small heads for an owl.

The ear tufts (false ears) found in some owls arise from the outer disk. They enhance the owl's broken-branch disguise when it perches in the open during the day and have nothing to do with

Fig. 8. Ear tufts that aid in concealment can arise in a variety of ways. Some species raise small groups of feathers selectively (*left,* Northern Pygmy-Owl), large feather groups, including parts of the facial disk (*center,* Northern Saw-whet), or specialized, enlarged feathers arising from the disk (*right,* Western Screech-Owl).

hearing. Tufts are sometimes also called "ears," as in Long-eared Owl *(Asio otus)*, or "horns," as in Great Horned Owl *(Bubo virginianus)*; they are neither.

Part of the facial disk, but usually distinct from it, are two curved fans, composed of bristlelike feathers (rictal bristles) that project forward and down from below the eyes and partly surround the beak. These can be raised and lowered and are tactile in function but may have additional uses; for example, those of a Snowy Owl are exceptionally thick and well developed (see fig. 105).

Continuing up the face from these fans are ridges of bulging small feathers that not only constitute the inner margins of the facial disk but also form the browlike arches between and above the eyes. Like the fans, they are often conspicuous and lighter in color than the other disk feathers and may function in visual signaling. These "eyebrows," too, are movable, as is the entire disk, by means of muscles: unique among birds, owls can therefore actually make faces. The outer edges of the disk can be seen to twitch periodically as the bird apparently scans its surroundings for sounds while it is dozing with closed eyes.

Fig. 9. Great Gray Owls can expose their beaks threateningly by raising and spreading their rictal bristles.

In some tuftless owls, such as the Spotted Owl *(Strix occidentalis)*, the upper right margin of the disk extends higher than the upper left so that it looks like a raised brow, imparting an arch expression to the owl's face. This asymmetry reflects a difference

Fig. 10. Owls can actually make faces by altering the shape of their facial disk. Barn Owl shown.

Fig. 11. The “raised eyebrow” (actually the right upper ruff) of the Northern Spotted Owl is the result of very asymmetrical ears.

in size or shape (or both) of the owls' ear openings. Owls with asymmetrical ears may have skulls that are distorted.

The base of an owl's skull, like that of other birds, bears a single condyle (knob) articulating with the vertebral column, rather like the ball pivot on a good tripod, on which the head can turn with ease (owls have a startling way of turning their heads not only almost upside down but what appears nearly clear around); humans and other mammals, by contrast, have two knobs, which limit head rotation. In addition, owls, like many other birds, have 14 cervical (neck) vertebrae, making the neck very flexible. An owl can turn its head 270 degrees from forward, a distinct advantage for an animal that has minimal eye movement. Hawks have similarly flexible necks, but lacking the broad flat faces and huge staring eyes of owls, they look less startling when the head is turned 180 degrees to the rear, which makes an owl's head appear to be mounted backward.

The hooked beak is heavy, even in small species, and sharp edged. The beak's curve drops rapidly, eliminating obstruction of the field of vision. A fleshy cere forms a saddle where it joins the skull, and partly surrounds the nostrils.

Senses

Smell and taste appear to be the only senses that are not exceptionally well developed in owls. Great Horned Owls even feed on skunks and have no problem with that mammal's familiar defense, and smaller species feed on insects that have all manner of repulsive chemical defenses. But owls have superb hearing, and their eyesight in low light is outstanding. The sense of touch, too, appears to be good.

Vision

Perhaps the most arresting features of an owl are its eyes, whether they are dark and soulful, like those of a Spotted Owl, or the blazing orange orbs of a Great Horned Owl. Although there is some variation between species in the relative size of these organs, the largest eyes may comprise as much as 32 percent of the skull weight, compared to 1 percent in humans (Korbel 1998).

Many predators, especially those feeding on fast-moving prey, greatly benefit from forward-facing eyes, a position that much improves three-dimensional vision and assessment of distance. Raptors that hunt by day (e.g., hawks) have an advantage in the lateral placement of their eyes, which they can rotate forward while still enjoying a very large field of vision to the sides and, to some extent, to the rear. For owls hunting in the near absence of light, however, lateral vision is of little value, and their almost immobile eyes have moved into a permanent, chiefly forward position to probe the darkness ahead. They have become elaborate and sophisticated traps for whatever little light is available, and the task of scanning the environment in most owls has been largely shifted to the ears.

Although forward facing, an owl's eyes diverge a bit so that the bird has a narrower stereoscopic range than we do, where the visual fields of the two eyes overlap. On the other hand, owls consequently have better peripheral vision, though not nearly as good as that of birds with laterally placed eyes, an extreme example being the long-billed, mud-probing Wilson's Snipe *(Gallinago delicata),* which can see better to the rear than to the front, a felicitous arrangement when early detection of predators is the chief concern. Forward-facing eyes, however, likely allow summation, the adding up of available light when such is scarce, and they provide binocular vision, which enhances depth perception and facilitates motion detection. Frontal eye placement also aids synchronization of the visual input with auditory information supplied by the equally forward-facing ears, an important ability when searching for prey under very low light conditions.

The main parts of a vertebrate's eye are analogous to those of a single-lens reflex camera. Light reflected by the environment enters the organ's chamber through the cornea (comparable to a transparent, flexible lens cover), and then passes through the pupil (the camera's aperture). Controlling the size of the pupil, and hence the amount of light that enters the eye, is the iris, the colored part of the eye (analogous to the camera's diaphragm). Next, the light continues through the lens and finally reaches the retina, the camera's film, where highly specialized nerve cells form an image that is passed on to the brain for recognition or interpretation.

The actual shape of birds' eyeballs is variable; some are roughly spherical, like a human's; others are flatter, as in waterfowl, for example; and those of diurnal raptors are conical. The distinctive great size of most owls' eyes is the result of the major enlargement of the cornea, pupil, and lens to admit maximum light under low-light conditions. These modifications have resulted in an eye shaped like a tube that flares out at one end so that it resembles a church bell. The mouth of that bell is covered by the enormous curved retina, which receives not only a very bright image but also a very large one: the eye's tubular shape allows for

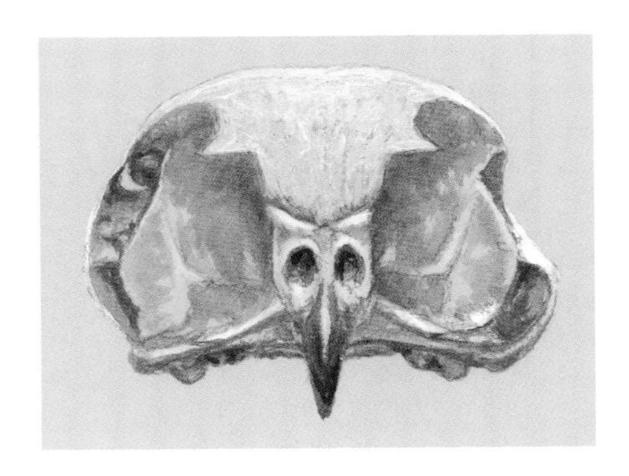

Fig. 12. Distorted skull of a Northern Saw-whet Owl resulting from the asymmetry of the skull's ear chambers.

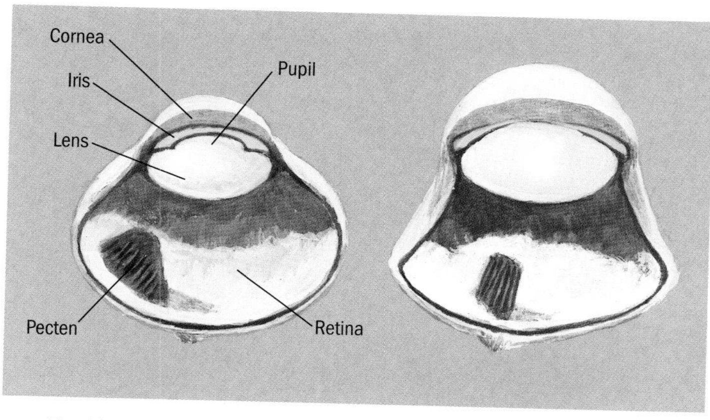

Fig. 13. Sagittal sections of the conical eye of a Northern Goshawk *(Accipiter gentilis) (left)* and the bell-shaped eye of a Great Gray Owl *(right)*. Found in reptiles and birds, the pecten's function is unknown. (Drawing after Korbel 1998.)

a substantial distance between the lens and the retina (the film), thereby enlarging the image, as a telephoto lens does when fitted to a camera.

The changes in shape and size of owls' eyes demand modified eye sockets, which in these birds are relatively shallow or even flat like a saucer. Lacking the protection and support of a deep surrounding cup, each eye has an internal sclerotic ring, a short, flaring tube of interlocking bony plates for added rigidity (smaller versions are found in other birds and most reptiles).

Although owls do not do well visually in utter darkness, the superiority of the light-gathering ability of their eyes becomes obvious when put in terms of the *f*-value of camera lenses: the Eurasian Tawny Owl's *(Strix aluco)* eye is *f*1.3, while a human's is *f*2.1 (Marks, Cannings, and Mikkola 1999). This difference is chiefly the result of the much greater size of an owl's pupil: it can dilate vastly more than a human's. Still, House Cats *(Felis silvestris)* have far better night vision.

Nocturnal birds, and in particular owls, have the reverse of the cone-to-rod ratio of diurnal birds. Rods (for black-and-white vision under low-light conditions) comprise 80 percent of the light receptors, the remainder being cones (for color vision), which,

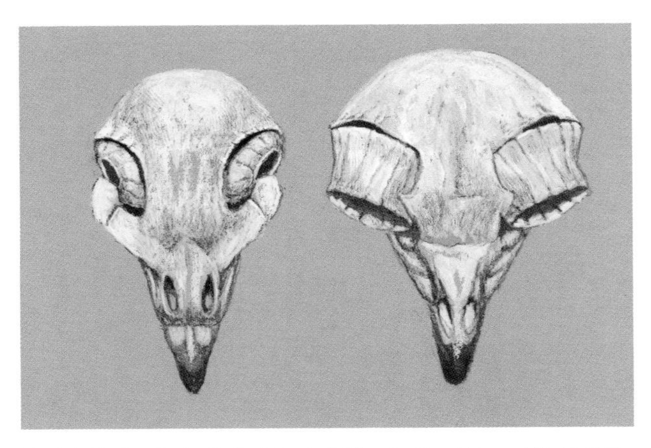

Fig. 14. Skulls of a Red-tailed Hawk *(left)* and a Great Horned Owl *(right)*. The hawk has prominent supraorbital ridges that shield the eyes, and the owl has enormous sclerotic rings to support the exposed eyes in their very shallow orbits.

however, require bright light to function. Owls also lack the capacity to perceive ultraviolet (UV) light, important to some other birds (Bowmaker and Martin 1978).

Although an owl obviously benefits from abundant rods, its visual acuity (sharpness of vision) is likely not all that great because acuity depends on numerous cones. However, the thick, bulging lens with excellent light-gathering power and the extra large size of the retina compensate for this problem by providing a very large image. Experiments with Great Horned Owls have demonstrated that because of its large number of rods, the retina of this species functions optimally in low-light conditions, but, because it also bears cones, it works under bright conditions as well (Ault and House 1987); contrary to popular opinion, owls can see perfectly well in daylight.

As in other birds, the pupil of these raptors can dilate or contract extremely fast because the surrounding iris in birds is operated by fast-action striated muscle instead of the much slower smooth muscle of a mammalian iris. Owls, like other animals, reduce their pupil size to cut back on the amount of light entering the eye; but they may also do so perhaps to improve the depth of

Fig. 15. The tubular eyes and maximally open pupils of a Great Horned Owl.

field perception prior to moving short distances, such as jumping to a new perch, as a photographer reduces the diaphragm diameter to achieve a greater depth of focus (Heinrich 1987). When owls oscillate or bob their heads or rotate their bodies in an exaggerated fashion and crane their necks as they look at a novel object, they are probably triangulating on it to judge the intervening distance.

Whereas a camera is focused by simply moving the lens forward and back, focusing the eye on an object is achieved by changing the shape of the lens (a process called accommodation) along with that of the cornea. Muscles that change the thickness of the lens are anchored to the inside of the sclerotic ring, which also prevents deformation of the eye's shape when these muscles contract.

Many owls have problems focusing on small objects close up. A captive young Western Screech-Owl *(Megascops kennicottii)* would pounce on a crumpled piece of paper tossed on the floor and, having missed it by just two or three inches, would back away and stand very tall to catch sight of the toy again and renew its attack. However, screech-owls and even much larger species catch insects and bats readily, even in the air, so their close-focus problems clearly are of little or no consequence.

Unlike other birds that close their eyes chiefly by raising the lower lids, owls do so by simultaneously lowering their upper lids as well, and they can seductively wink or blink with their upper lids like a practiced courtesan. An interesting feature of

Fig. 16 (right). Like humans, and unlike most other birds, owls blink by lowering their upper eyelids, as seen in this Long-eared Owl.

Fig. 17 (below). The nictitating membranes (third eyelids) are partway drawn across the eyes of this Western Screech-Owl as it prepares to bathe.

birds (and other vertebrates) is the nictitating membrane, a third eyelid (lined with brush-tipped cells) that sweeps across the eye diagonally to clean and moisten its surface (the tiny pink fold in the inside corner of a human's eye is an evolutionary vestige, a reminder of our distant reptilian ancestry). In owls, this structure is semitransparent and also serves to protect the eye, for example during bathing and while seizing prey.

Hearing

Remarkable as owls' eyes are, they are easily matched and even surpassed in performance by their ears. Hearing enables owls to locate prey in what amounts to complete darkness, and to that end, their ears are exquisitely modified. Upon discovering a perched Short-eared Owl *(Asio flammeus)* the length of a football field away, it is instructive to pucker up and suck in one's breath sharply. This will produce a sound similar to the soft squeak of a mouse and cause the owl's head to whip around at once to face the squeaker. A captive Northern Saw-whet Owl *(Aegolius acadicus)* busy eating in a noisy room, with people talking and the television turned on, instantly stopped and stared fixedly at a distant corner of the ceiling whence came the faint rasping sounds of a mouse gnawing. Probably all owls listen very selectively and screen out sounds of no interest: the rattling of wind-blown leaves, the shuffling of an opossum in the litter, the creaking of limbs and trunks—all these need to be filtered out.

Barn Owls, which have been subjects of a great many experiments, pounce on rustling bits of paper towed by mice and not on the rodents, showing that they do not rely on smell or infrared radiation to find prey (Konishi 1973). Blocking one ear prevents the owl from pinpointing a sound, and removal of most of the facial disk feathers results in the owl striking far short of its target, showing that the disk apparently functions as a sound-amplifying and focusing system. However, if one ear is plugged in a young owl that is still growing, it can compensate for this loss (Knudsen 1981; Konishi 1973, 1983, 1993).

Because space is three dimensional, an owl must pinpoint the prey's location in terms of distance as well as vertically and horizontally. Sound produced by the prey, such as the rustling of a mouse, offers two clues as to its whereabouts. One is the time interval between the arrival of the sound at the right ear and the left ear, which furnishes the horizontal coordinate. A Barn Owl, for example, can distinguish arrival intervals as short as 10 millionths of a second (Konishi 1993). The great distance between the ears in owls is obviously critical.

Another directional cue is the sound's intensity, or force of the sound waves, which is strongest in the ear that first receives it. In the Barn Owl, the left ear opening is located above eye level but points downward, while the right ear is below and points up,

Fig. 18. Left and right ear openings of a Barn Owl, showing asymmetry and, on the left ear, the ear flap (operculum, arrow) that catches sounds coming from *behind* the head.

receiving sound more readily from above. This ingenious arrangement supplies the vertical coordinate (Konishi 1993), derived from the sound's intensity. Finally, very low pitched sounds consisting of a series of long, low waves may convey directional information through phase differences, a method not used by Barn Owls.

Besides asymmetrically placed ears, some species have ear openings of different sizes or shapes for differential sound reception, or they may have flaps (opercula) to shut out the sound or to direct it, funnel-like, into the external auditory canal, including sound coming from *behind* the head—like human ears mounted backward, in front of the ear opening rather than behind. Some species have combinations of these features, or they may be able to change the shape of the ear openings. Overall, tytonids have a relatively small ear opening with a large operculum, whereas in strigids the ear shape is more varied. In the Northern Saw-whet Owl, both ear openings are enormous (see fig. 131) and of equal size, but the ear chambers of the skull within the head differ greatly in size and location.

To summarize, in Barn Owls, information about the horizontal location of a sound source is provided by arrival intervals at the ears, and vertical information is supplied by the sound's intensity as received differentially by the asymmetrical ears. It is not difficult to visualize the joining of these two methods as the crosshairs of a rifle scope—with a mouse in the middle. With the possible exception of the diurnal sight-hunting species, all owls likely use this bimodal system of locating prey in darkness. How they assess distance, however, is not well understood.

The span of the hearing range, from high-pitched squeaks to lowest rumbles, varies between owl species. Generally, owls' ranges resemble ours; they are capable of hearing high-pitched sounds as well as we can but do less well in the deepest basso profundo ranges. However, when the *volume* of lower sounds is substantially reduced, owls come out ahead handily, with some performing 10 times better than humans do, perceiving sounds that we cannot hear at all. This ability to hear noises where we hear nothing enables some to catch prey in total darkness and in broad daylight under grass, soil, or snow as deep as 45 cm (18 in) (Marks, Cannings, and Mikkola 1999).

Input of sound results in the firing of certain nerve cells in the brain that are specialized to respond only to sounds coming from specific areas around the owl; in this way, a map is formed by these space-specific neurons that is a copy of the owl's environment, a map that, like a cruise missile's computer, guides the owl unerringly to the rustling mouse. Such auditory maps in the brain are first generated with the aid of vision. During development, both auditory and visual input have to coincide; when this occurs, a "learn" signal is turned on in the brain, wiring neurons into a three-dimensional map configuration (Haessly et al. 1995). If a young owl is fitted with special glasses that make the bird see at some angle to the side instead of straight ahead, the owl forms an additional auditory map, based on this distorted information, which again provides an excellent and accurate navigation system (Hyde and Knudsen 2002). Auditory maps are continually updated and modified as needed: the owl's brain has a "gate," which, though normally closed, allows visual information to enter the auditory area to make the map current. There is even a mechanism to point up errors (Gutfreund et al. 2002).

Touch

The sense of touch does not appear to have received much attention in owl studies. Most owl students agree that the rictal bristles surrounding the beak are of a tactile nature, for owls nearly always close their eyes when they reach for something with their

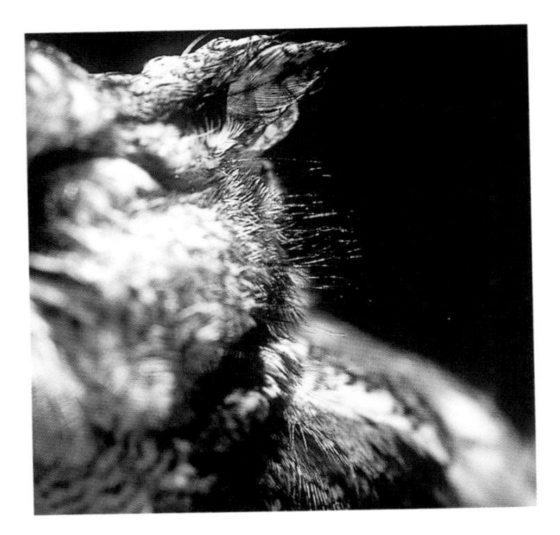

Fig. 19. The fine filoplume-like feathers projecting from the side of the head of a Western Screech-Owl, possibly tactile or sound-muffling in function.

beaks and seem to feel their way to the prey's head or to the head of a fellow owl for allopreening. Because owls are also decidedly farsighted, they may have to rely entirely on touch at close quarters. A Northern Saw-whet mistook its own dropping on the floor for a food item, shuffling over it with eyes closed, blindly feeling with its rictal bristles and biting at it. Discovering its error, the owl violently shook its head to dislodge the material from its beak.

Perhaps the filoplumes (or filoplumelike feathers) projecting beyond the head's contour feathers serve to assess the size of openings as do a mammal's whiskers. However, these fine thread-like structures are especially obvious in the Snowy Owl, which does not enter holes, and the toes of Barn Owls also bear bristles, the function of which does not readily suggest itself.

Voice

Because owls cannot communicate in the dark with colors, postures, and feather displays like day-active birds do (although they do flash their stark white signal patches), they have evolved a rich repertoire of vocalizations for social signaling that enables them to function fully at night.

The nocturnal hoots of owls are familiar sounds of the rural and sometimes suburban West. They are often sung, surprisingly, with beaks closed, and they serve to proclaim territories and attract mates. In addition to these advertising (or primary) songs, owls are able to produce a great variety of shrieks, yelps, barks, whines, whinnies, howls, whistles, and chatters. They also scream, screech, moan, purr, chuckle, bleat, yowl, cackle, hiss, and tick like a grandfather clock. Oddly, screech-owls do not screech (except during defense of their young), but Barn Owls do, as was observed by Shakespeare. Some of these vocalizations serve as warning signals of varying intensity to fellow owls (Sproat and Ritchison 1994). One early November, three or perhaps four Western Screech-Owls within earshot of one another produced a great variety of calls in a short period. Overall, the full vocal repertoires of the different owl species are not well known at all, nor are all the meanings of all the various calls. Much is yet to be learned.

A captive female Western Screech-Owl invariably responded to her keeper's sneeze with a drawn-out wail resembling that of a woman in great distress. One scream, presumably emitted by an unseen Long-eared Owl, nearly brought on a cardiac infarct in the then young and innocent author during a moonless midnight walk in the Sierras in search of Flammulated Owls *(Otus flammeolus)*. Barn Owls, when handled or even approached too closely, can let out blood-curdling shrieks at such volume as to turn heads a hundred yards away, and Northern Saw-whets have a call that eerily resembles the whine, trailing off rapidly at the end, of a ricocheting bullet.

In some owls, and perhaps in all, the male's singing voice is lower in pitch than that of the female, his smaller size notwithstanding; this is very noticeable in Great Horned Owls, for example. Some calls, such as food delivery calls, are given by males only. Identifying the sex of a calling owl by its voice is useful to field biologists.

Juvenile owls stay in contact with each other and with their parents through special calls and vice versa. A fledgling Northern Pygmy-Owl was visibly activated by hearing its hidden nestmate's begging calls.

Apparently, all owls snap their bills (a mechanical sound), often repeatedly (and even to the point of cutting their tongues) when they are excited, afraid, or feeling aggressive.

Skeleton and Legs

An owl's skeleton is, like that of most other birds, extraordinarily lightweight and compact, forming an oval streamlined box into which are packed the internal organs, with limbs and muscles relating to body movement attached to the outside. To reduce weight, the longer bones are hollow and thin walled, with fine internal struts that provide strength. The relatively short sternum and keel of the majority of owls suggest only moderate flight muscle development; most owls fly with some speed, but they do not approach the velocity of some diurnal raptors. Some, however, may fly for long periods or migrate great distances. Although owls' necks may seem especially flexible, they do not differ in this respect from other birds, and, as in these, the rest of their vertebral column is largely fused to provide a solid airframe, with but a single moveable joint between the back and the pelvis.

Most people familiar with owls are of the opinion that their bulky-looking bodies are really a sham, that all those fluffy feathers disguise a weakling's body. It is true that the plumage of most

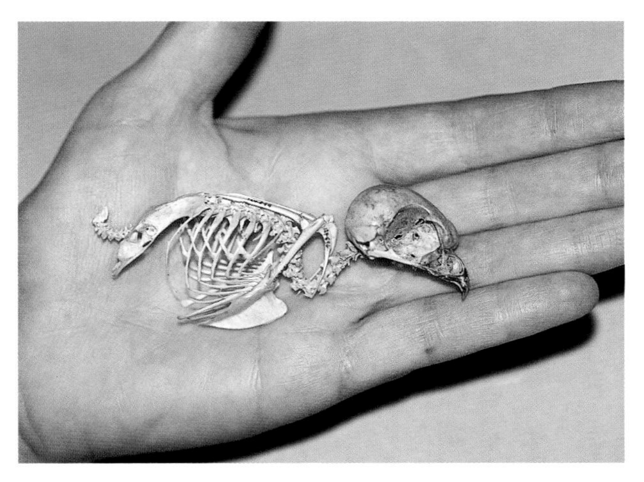

Fig. 20. The compact, boxlike skeleton of a Northern Pygmy-Owl, typical of all birds that fly.

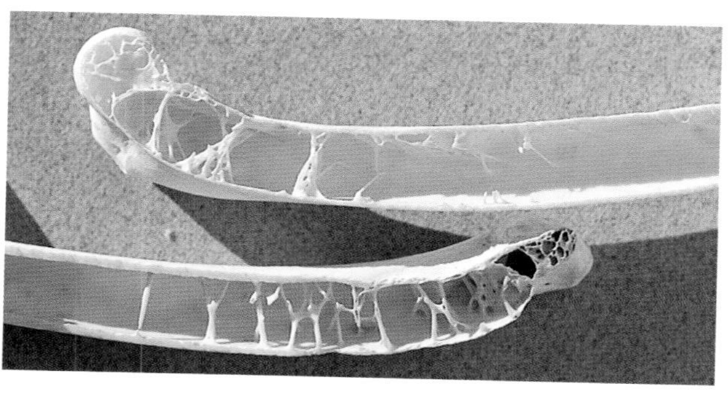

Fig. 21. The thin-walled, lightweight long bones of birds, such as these from a Great Horned Owl, have internal struts (trabeculae) to strengthen them.

owls suggests a much greater body size and mass and that no owl has the chest musculature of a swift falcon (though one, the Snowy, sometimes hunts like one), but even a little screech-owl held in the hand discloses that its small, compact body is equipped with legs that are extraordinarily thick and muscular, like the arms of a bantam-weight boxer. The Great Horned Owl, which might briefly be mistaken for a Red-tailed Hawk *(Buteo jamaicensis)* as it flaps through the trees in uncertain light, in fact weighs nearly half again as much as the hawk that appears so similar in size.

Although the legs may be longer in some species, as in Barn and Burrowing *(Athene cunicularia)* owls, those of nearly all of North America's owls are of very similar sturdy build despite the great range in size, the one exception perhaps being the more delicate Flammulated Owl. Stout in appearance, the legs are usually equipped with relatively thick and obviously powerful toes. In some species, these are long as well, suggesting that the owner may include birds in its diet.

Like most other birds, an owl's foot has three toes that face forward and one (designated as toe 1) that extends to the rear (the latter being the equivalent of the human big toe rotated backward; the little toe has disappeared). Such an arrangement provides the bird with the ability to grasp, which in owls is further enhanced by facultative zygodactyly, the ability to rotate at will the

Fig. 22. Typical "stiff-arm" attack of an owl. The feet, with spread toes, are brought together as a catcher's mitt. Long-eared Owl shown.

outer forward toe (toe 4) to the rear. This talent is also seen in the Osprey *(Pandion haliaetus)* and the African touracos (a number of other bird groups, such as parrots, are also zygodactyl but do not have the forward alternative). An owl therefore has the option of deploying the usual avian three-forward-one-back toe arrangement or switching to the two-by-two configuration, which is probably useful when carrying heavy prey and when seizing small mammals, as well as for perching. Unlike hawks, owls fly with clenched feet, even as they approach prey. The eight toes spread, just before contact, into a symmetrical configuration to cover as large an area as possible (Payne 1962). When not attacking, owls land softly on the ground but frequently alight heavily on branches and similar perches, often with an audible thump, perhaps to ensure a solid landing in low light.

The undersides of the toes and the soles of owls' feet are fleshy and soft to the touch, like an old leather valise. In most owls, they bear numerous round, somewhat flattened pads, and they are entirely and densely covered with tiny nipplelike papillae that look like they are perfect for clamping onto fur. The toes terminate in

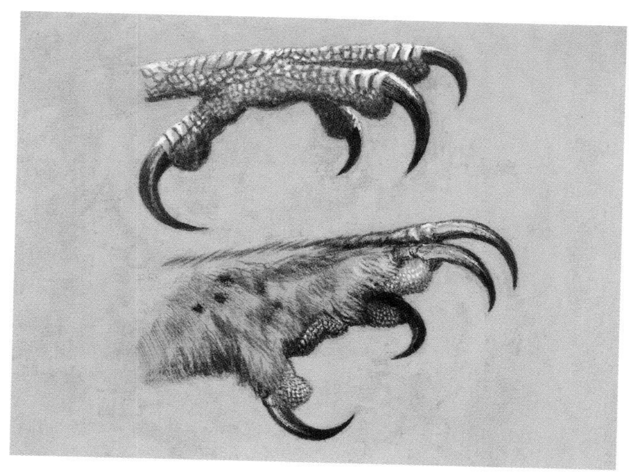

Fig. 23. The longest talons that aid in killing derive from different toes in hawks and strigid owls: Red-tailed Hawk *(upper)*, Great Horned Owl *(lower)*. Tytonid talons are more hawklike.

exceptionally sharp talons that can produce exquisite pain in a captured owl's careless handler, who presently also becomes aware of the amazing strength of the grip of even a small species.

Whereas in hawks and eagles it is the rear and inner toes (toes 1 and 2) that bear the longest talons, used in killing prey, in owls the rear talon (hallux) is small and is in fact sometimes the smallest claw, so that most stabbing action is left to some of the longer forward talons. Owls, however, chiefly kill by biting into the neck, or more often, the skull of the prey.

Displaying the great strength of its legs and feet, an injured Northern Pygmy-Owl was able to maintain its body in a perfectly horizontal position while clinging to its handler's finger with nearly straight legs (K. Richman, pers. comm. 2005). A Great Horned Owl demonstrated its fitness and resolve in the days when free-range chickens were the norm. During one of its regular raids on a farm, its left foot was caught in a steel jaw trap, which it carried off, anchor chain and all. It returned for an additional three chickens on subsequent nights, chain clanging, when, on its fourth visit, its trap was caught in a second and larger one.

Closer examination revealed that this tough bird had also been carrying a shotgun pellet in its left eye socket, rendering it blind in that eye (Allert 1928).

Wings

With feathers, skin, and flesh stripped away, it becomes clear that a bird's wing is actually an arm and hand in disguise, the bones readily recognizable (the "buffalo wing" sold in restaurants is obviously a chicken's lower arm containing radius and ulna) except for the wrist and hand where, for weight and strength considerations, bones are reduced in number with the remaining ones largely fused. Only the thumb persists as a separate finger, giving rise to the alula, a group of feathers related to braking while landing and to steering during certain maneuvers. Its derivation from the first digit is nicely illustrated in a newly hatched chick, for example, which bears a claw ("nail") at the tip of the appendage.

A fully feathered wing forms an airfoil; its leading edge bulges, then tapers, with a fore-to-aft curvature to a thin trailing edge. It is this cross-sectional shape, copied and easily seen in an airplane's wing, that provides the lift that enables a bird to fly.

Compared to the wings of diurnal raptors, those of owls at first glance seem to show remarkably little diversity in shape. However, on closer inspection, substantial differences become obvious. Owls that live in woodlands and forests have broad, rounded wings (the wing of a Western Screech-Owl is reminiscent of a ping-pong paddle), whereas the wings of species that forage over open country are narrower, longer, and somewhat pointed. The highly migratory Flammulated Owl's wing has a conspicuously long hand, and the sometimes songbird-catching Northern Pygmy-Owl has the pointed wings of a speedster. Still, owls' wings provide a large surface area that carries a relatively low body weight, allowing these birds to fly slowly and make quick turns without stalling.

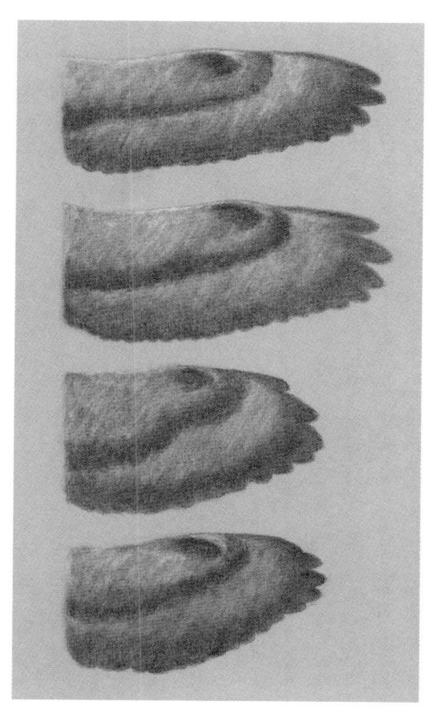

Fig. 24. Adaptations of owl wings *(from top):* open country, long-range, long and narrow (Barn Owl); combination woodland–open country, moderately long and broad (Great Horned Owl); woodland-forest, short, high lift (Northern Saw-whet); woodland-forest, short, high lift, pointed (Northern Pygmy-Owl).

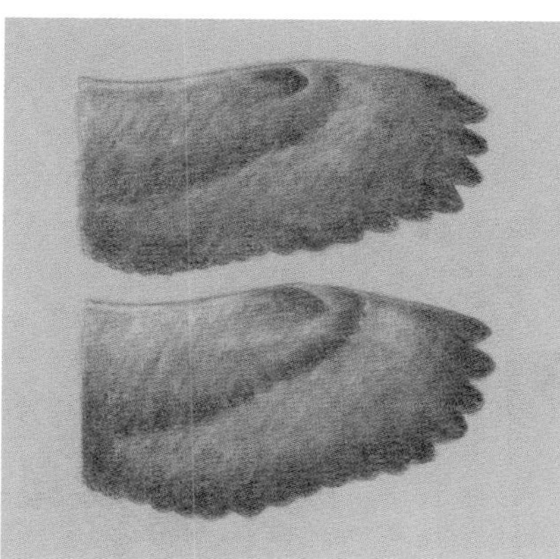

Fig. 25. Wing of a migratory forest owl (*upper,* Flammulated Owl), showing longer, narrower hand than that of a nonmigratory woodland owl (*lower,* Western Screech-Owl).

Fig 26. Owl topography. Long-eared Owl shown.

Feathers

The overall shape of a bird is molded by its body (contour) feathers, including those of the wings and tail. The long flight feathers of a bird's wing (the remiges) are divided into two groups. The secondaries, which arise from the lower arm, keep the bird airborne. The primaries, attached to the hand, not only carry out the same function but also provide forward propulsion. The flight feathers of the tail (the rectrices), variable in length depending on the species, can be spread to provide braking and additional surface during slow-speed landing and, to a lesser extent in owls, to steer. Most of an owl's facial disk is made up of several types of highly modified contour feathers.

Silent Flight

Silent flight is one of the hallmarks of the great majority of owls, and the upper surfaces of the flight feathers and even coverts are endowed with a velvetlike nap that absorbs the sound of the

Fig. 27. Primaries of a Great Horned Owl (left) and a Red-tailed Hawk. The owl's feather is covered with a thick pile.

feathers rubbing together. A similar pile is present on the remiges of two diurnal raptors: the Northern Harrier *(Circus cyaneus)*, remarkable for having an owlish facial disk, and the White-tailed Kite *(Elanus leucurus)*, a hawk that, for a day-hunting species, has unusually large eyes, which suggest foraging in low light (a close relative in Australia hunts by night).

Despite this covering, owl flight feathers are still so thin that you can read a newspaper through the unbarred portions. So light are they that a Great Gray Owl's *(Strix nebulosa)* 30-cm (12-in.) tail feather floats to the ground when dropped (Nero 1980); a similar-sized hawk's rectrix spirals down. In addition, the outer primaries of most owls bear comblike leading margins that also

Fig. 28. Despite the pile, owl flight feathers are so lightweight and thin that you can read a newspaper through the unpigmented parts. Barn Owl primaries shown.

muffle wing sound, as do the finely tattered trailing edges of the primaries and, often, secondaries. The fringed exhaust nozzle of a jet engine, which reduces noise, is a direct copy of such an owl feather.

Noise reduction is essential not only to keep prey from hearing an owl's approach but especially to prevent extraneous

Fig. 29. The comblike margin of the outermost primary of a Great Horned Owl *(top of photo).*

sounds produced by flight from swamping the raptor's acoustic receptors during an attack guided by the rustling of the quarry. Owls are drop shaped, with the head forming the wide end of the drop. In flight, this thickly feathered protuberance and, in particular, feathers of the facial disk (which points forward) may in part function in the elimination of airstream noise near the ears, much like a fuzzy microphone cover reducing wind noise during an outdoor recording session.

Interestingly, the comblike leading edge of primaries is nearly absent in some species, such as the Northern Pygmy-Owl, where these combs are rudimentary or, more likely, vestigial; this diurnal owl has a noisy, whirring flight reminiscent of that of a California Quail *(Callipepla californica),* having no need to muffle sound: its prey probably considers the approaching raptor just another songbird because of its small size.

Thermoregulation

The soft, layered body feathers of most species (only the Snowy and Northern Hawk *[Surnia ulula]* owls have stiffer feathers, to deal with Arctic winds [Duncan 2003]) are remarkably downy toward their bases as well, and not only deaden airstream sound in flight but also provide excellent (albeit at times unwanted) insulation in the majority of owls. Most have feathered legs and, often, feathered or partly feathered toes. While some biologists think these "socks" reduce injuries from the bites of prey animals or muffle airstream noise in flight, they seem actually related to retaining body heat: owl species living in colder northern climates usually have more densely feathered legs and toes, and even within a species such as the Barred Owl *(Strix varia)*, populations in the northern extremes of its range have dense feathering on both legs and toes, whereas their brethren in balmy Florida wear leggings but go barefoot (Marks, Cannings, and Mikkola 1999).

Owls, like other birds, also puff out their feathers when cold, to increase insulation; this action enlarges the airspace that traps heat given off by the body. Still, maintaining the body at a temperature that is both comfortable and physiologically advantageous can be problematic for owls. Because most species are so well insulated by their feathers, cold weather is of little consequence, but heat can be troublesome. Western Screech-Owls hang out of their nest boxes (and presumably thin-walled tree cavities) when it gets too hot inside. The Northern Spotted Owl *(Strix occidentalis caurina)* exposes the soles of its feet, stands tall to expose its legs, and raises breast and back feathers to facilitate heat loss (Barrows and Barrows 1978); one reason this owl requires an ancient multilayered forest is that it provides a great variety of thermal microclimates for optimal roost sites (Askins 2000). Neonate owl chicks are virtually incapable of thermoregulation.

Owls (and many other species of birds) also use gular fluttering for cooling off. Normal breathing results in negligible loss of water, but because water absorbs much more heat than air, substantial heat can be removed from a temperature-stressed body by increasing this respiratory water loss. Gular flutter is the rapid vibration of the upper throat air passages, with the beak hanging open, which pumps out heated water vapor. The Spotted Owl is more heat sensitive than the Great Horned Owl. Gular flutter was

Fig. 30. Although screech-owls often look out of their nest boxes, they project more of their bodies if the box gets too warm. Western Screech-Owl shown.

observed in Mexican Spotted Owls at 30 °C (86 °F) but not in Great Horned Owls *(Strix occidentalis lucida)* until the temperature reached 37 °C (98.6 °F) (Balda 1993).

Plumage

Resilient as they are, feathers are dead structures that abrade and wear out and therefore must be shed and regrown periodically. In many birds, all body and flight feathers are replaced annually in one or more molts. Raptors' flight feathers, important for hunting, may undergo a very protracted molt that may go on, at intervals, for three years or more before all feathers are replaced. The food supply strongly influences the rate of molting: low food availability greatly slows down the process. However, some small owls molt all their tail feathers at once so that they are temporarily tailless, indicating that these appendages are likely of no great importance in their lives. Owls' tails are mostly short, almost comically so in some species.

The colors of owl feathers tend to be subdued; grays, tans, and black browns predominate, with some species (or forms within species) showing variable amounts of rufous brown or ocher and rust. These colors, often arranged in subtle patterns or bold lines to break up the bird's outline, make owls nearly invisible against bark and other tree parts, or in weedy fields.

Fig. 31. The conspicuous white eyebrows of juvenile Northern Saw-whets probably make it easier for the parents to locate and identify them in the dark once the youngsters have fledged.

Highlights, especially on the throat and around the face, are usually white and are very visible in poor light. The Great Horned Owl, for example, has a bright white throat patch that is particularly prominent when the bird is calling. In the Eurasian Eagle-Owl *(Bubo bubo),* a closely related and similarly colored species, it was shown that this "badge" varied seasonally and differed between the sexes in brightness; females had higher brightness values than males, and the reflectance of the throat patch was positively correlated with the size of the female (but not the male) (Penteriani et al. 2006).

Some owl species have distinct color phases (morphs) where individuals are differently colored from their conspecifics while more or less retaining the usual pattern. For example, the Eastern Screech-Owl *(Megascops asio)* has a strikingly reddish form besides the more widespread gray type, as well as intermediates, and the Western Screech-Owl has both a gray and a brown phase. At present, no ecological explanation for these variations is entirely satisfactory. It is possible that they are no more meaningful from an adaptive standpoint than the red-haired people scattered in human populations. However, red screech-owls are said to die in greater numbers in winter than the gray form, which seems, therefore, more cold adapted (Moser and Henry 1976); but why then are red birds so uncommon in the warm southeast? They seem certainly preadapted to life in urban environments, which are generally warmer and more stable; and the gray form is more fit to survive periodic climate changes that bring cooler weather (Gehlbach 1994a).

Some hatchling owls are nearly naked but quickly sprout a short, variably dense coat of usually white down, and others are born with it. Typically, a new covering appears within one to three weeks (depending on the owl's size), a second downlike coat, the juvenal plumage, that is normally subtly colored and often variably barred. At this stage, young owls leave the nest (prefledge) and cannot yet (or just barely) fly, although their flight feathers have made their appearance, along with feathers of the back and, in some, of the flanks. Juvenal plumage is soon replaced by the so-called Basic I plumage, similar to an adult's. By the age of four months or so, most owls resemble their parents, although their coloring may be drabber.

Hygiene

An owl's feathers are part of its tools of the trade, and because they are frequently highly specialized, their maintenance is of primary importance. In addition to the often complex flight feathers, the body feathers for the most part are exceedingly finely structured and, bearing long, downy barbs, become easily matted. To top it off, owls frequently carry an extraordinarily high load of ectoparasites that suck blood or feed on the feathers.

No surprise then that an owl allots much time to preening. The bird draws each feather individually through its beak, from

Fig. 32. Burrowing Owls at their nest hole. Preening is often followed by yawns. Stretching one wing includes simultaneously stretching the leg of the same side.

base to tip. The long primaries, subjected to this procedure, bend almost alarmingly and audibly snap as they slip out of the beak.

Owls also preen the feathers of their legs and feet and nibble on their toes to remove dried blood and other debris; they are generally more effective at this task than diurnal raptors, the feet of which are often soiled with blood. However, some are less successful at cleaning their faces, and although like hawks, they also wipe their beaks on branches or the ground to clean them, it is not uncommon to see Northern Pygmy-Owls, for example, with blood about their beaks. Owls use their talons to preen the head feathers (unreachable by the beak), scratching much like a dog scratches an ear. In Barn Owls, the combs on the middle talons come into play here and are themselves cleaned by the bird's beak.

To shed loosened debris, preening sessions are punctuated by "rousing," the raising and vigorous shaking of the plumage (in large owls an impressive procedure that resembles someone vigorously shaking wet clothes), and by reaching with the beak to the uropygial gland above the tail to pick up its secretions, which are then spread on the feathers with the beak, except for those of the head, which are rubbed directly over the gland. The oil produced by that gland contains a precursor to vitamin D that, when

Fig. 33. To dislodge loose feathers and debris and realign the plumage, birds fluff up and vigorously shake themselves. Northern Pygmy-Owl shown.

Fig. 34. The preen (uropygial) gland at the base of the tail is stripped with the beak, its oil then spread on the bird's feathers, where some of it is converted by sunlight into vitamin D. Burrowing Owl shown.

Fig. 35. Allopreening is an important activity among owls. It strengthens pair-bonds and the bonds between adults and their young. Burrowing Owls shown.

exposed to sunlight, is converted into that essential nutrient, ingested when the bird preens.

Paired owls, and sometimes siblings or parents and offspring, preen one another about the face and head, an activity called allopreening ("allo" means "other"), which is poorly understood. It may be mutual cleaning, or it may serve to reduce aggression or function in sex recognition and pair-bond maintenance. Certainly owls engaged in allopreening look like the epitome of a nuzzling couple, their eyes tenderly closed as they nibble one another. Captive owls raised from chicks commonly preen their keepers' hair (with the human returning the favor by scratching the owl's head, which offers its dome with blissfully closed eyes). In the Great Gray Owl, the drive to allopreen is so strongly developed that a freshly caught individual may preen its captor's head—even if the owl is badly injured (Nero 1980).

Probably all North American owls bathe. Burrowing Owls

Fig. 36. Some owls look like accident victims when they sunbathe. Burrowing Owl shown.

Fig. 37. This Northern Saw-whet juvenile is using a floor lamp for "sun" bathing, mistaking the light for a patch of sun.

excitedly run about in the rain, and Western Screech-Owls and Flammulated Owls nearly submerge themselves with great enthusiasm. All deliberately wipe their beaks afterward. At least some of the species also sunbathe. Great Horned Owls bathe in fine sand, and their elongate scrapes, often with a few molted feathers, can be found on woodland dirt roads (B. Mahall, pers. comm. 2006). Sandbathing and dustbathing have also been observed in Long-eared Owls (Nuijen 1992). These dry baths kill parasitic arthropods; dust particles destroy the exoskeletons of such pests through friction between the plates of the exoskeleton. The boreal Northern Hawk Owl has been observed bathing in snow (Cade 1952).

Parasites

Anyone who has handled a wild-caught or road-killed owl is probably familiar with odd looking, very flat bodied flies that scuttle at speed from under the feathers; they may slip back equally fast into their hiding places, or scurry onto the human's hands or fly off. These insects are members of the family Hippoboscidae, blood-sucking flies that in their various incarnations can also be found on deer, pigeons, grouse, and other hosts, humans not among them. Nevertheless, because of their Quasimodo-like, crabbish movements, most people find these flies creepy; they also transmit malaria to their avian hosts (Powell and Hogue 1979).

In addition to hippoboscids, owls are also parasitized by other flies (Diptera); the maggots of the family Calliphoridae, for example, suck blood from their young. Fleas (Siphonaptera) and bird lice (Mallophaga) are also found on owls, along with mites of various families that feed, for instance, on feather fragments, lipid secretions, tissue fluids, and even fungi, bacteria, and algae living on the owl. Two thousand quill mites were collected from one Great Horned Owl, where they fed on the medullae (cores) of feather quills. Because owls dig out certain mites with their beaks, the pests can cause extensive mange (Philips 2000). Finally, there are also protozoans that swim in owls' bloodstreams and worms that writhe in their intestines and air sacs. There is no shortage of freeloaders.

Deconstructing a Mouse

Digestive System

Feeding on easily digested food, a carnivore such as an owl has a simple digestive system that is, compared to that of a plant eater, short. Interestingly, owls that hunt by active flight have a lighter digestive system than those that perch-hunt (Barton and Houston 1996); weight reduction benefits an aerial hunter.

The stomach, as that of other birds, consists of two chambers, the proventriculus, which adds hydrochloric acid and protein-digesting enzymes to the food, and the ventriculus (gizzard), where food is mixed with these digestive agents and where an ingested mouse is quickly separated from its fur and most of its bones. The ventriculus is firmly anchored to the body wall, perhaps to keep it from shifting about in flight when full. The outlet to the intestine (by way of the pyloric sphincter) is located on top and is minute, preventing passage of all larger solids; only a soup of digested flesh and enzymes can pass. Production of this soup starts within a very few minutes after ingesting the food item, even if it is covered with hair.

In the course of evolution, birds have lost their teeth, which are relatively heavy structures and would make them front weighted in flight (though they still have the genes for making teeth). In the absence of dentures, a chicken or a sparrow uses swallowed grit and stones to grind up seeds and other vegetable matter in its thick-walled, muscular gizzard (the eminently edible giblet), located in the belly (the center of gravity). Because meat is so easily digested, carnivorous birds lack this "mill," and their gizzard is a simple soft pouch for temporarily storing food. The gizzard also acts as a sort of trash compactor that combines the hair, bones, and other indigestible materials into an oblong bolus (called a pellet) that is then coated with mucus and regurgitated as waste.

Unlike other avian meat-eaters (such as hawks and falcons), owls lack a crop, an internal storage sack that allows the former to carry much more food than the stomach alone can hold, making it worth their while (measured in terms of energy expenditure) to kill very large prey. The absence of a crop in owls is a real inconvenience for them. For example, a Northern Saw-whet Owl must eat a large mouse in installments, whereas an American Kestrel *(Falco sparverius)* of the same weight as the owl would easily pack away the rodent in one sitting. This means that the owl must either store the remainder of its meal for later consumption or else hold it in its foot (sometimes like an ice cream cone) for the next several hours until the first course has been digested and its remains in the form of a pellet coughed up. If the first part of the meal contains little pellet-forming material, such as feathers or fur, the owl may however feed again on its leftovers after a relatively short interval without first ejecting a pellet.

Three captive Northern Saw-whet Owls, fed a diet of laboratory mice *(Mus musculus),* ate the head and one or both front legs first. After a pause of four or five hours, a pellet was produced and the rest of the mouse eaten, minus the stomach and, sometimes, parts of the intestine. A second pellet was cast during the next nine hours, containing chiefly bones from the rear half of the mouse. Small mice were eaten in one sitting, though never swallowed whole in the manner of larger owls (Collins 1963). The big owls sometimes kill extraordinarily large (for their body size) prey and only eat part of it; they may then return to it for additional meals—if it has not been scavenged.

Owls produce generally one pellet per meal; if several mice are caught at intervals during the night, an owl may produce multiple pellets. The interval between eating and ejecting a pellet varies with the size of the meal, what time it was consumed, and other factors. A captive Short-eared Owl produced a pellet each time the owl was offered a dead mouse at intervals of several hours throughout the night. Some of these pellets contained considerable amounts of undigested muscle, indicating that the sight of prey is a powerful stimulus to make room for more food, at least in this species (Page and Whitacre 1975).

Four captive Great Horned Owls did not need to cough up a pellet before eating another meal, and in fact, as pellet weights indicated, digestion was more thorough when the earlier meal's pellet was retained (Duke et al. 1993). Apparently, the size of the owl and that of the food item can influence pellet production.

Two biologists, curious about how a wad of hair soaked in hydrochloric acid could safely pass up an owl's delicate gullet without burning it, sampled a Great Gray Owl's freshly produced pellet. One licked the mucus coating and reported it sweet to the taste. His colleague, unaware that the other's test involved no more than licking, bit into it and flinched from the extremely bitter taste of the interior of the wad (Duncan 2003). Oddly, in the Great Horned Owl, stomach juices are acid in young birds but neutral in older ones (Grimm and Whitehouse 1963).

Pellet Analysis

The ground under some owl roosts is at times littered with pellets in various stages of decomposition; they provide the naturalist with the splendid opportunity of sampling the local small mammal fauna, because pellets commonly contain skeletal elements and entire or partial skulls of mice and other vertebrates, providing diagnostic evidence of the various species of an area. In fact, a new species of gerbil was identified from its skull found in the pellet of an owl (Schlitter 1973).

Owl pellets can usually be readily told apart from those of hawks, which normally have few if any bones. Only the White-tailed Kite is more like owls in its failure to digest much bone, but its pendant-shaped pellet is very distinctive. The pellets of the Short-eared Owl and the Northern Harrier, which may live in the same habitats, can be distinguished by their lengths and bone contents (Holt et al. 1987). However, it is often impossible to tell apart the pellets of different owl species, even with the aid of photos, unless there is a great divergence in size between the local kinds of owls. Even then, the pellets of a given species are not consistently of the same size.

Besides the remains of rodents and other small mammals, owl pellets contain indigestible parts of other prey, such as the chitinous exoskeletons of insects (the parts that crunch underfoot), bird bones, feathers, beaks, and feet, and fish bones. Incidentally ingested leaves and plant fibers also show up, along with plant material (seeds, for example) and stones taken in by prey. Burrowing Owl pellets may be entirely composed of sand or may contain sand or glass accidentally swallowed while modifying ground squirrel burrows with their beaks, as well as horse or cow dung, which these birds use to line their nests (Martin 1971) or gather to attract beetles, which they can then eat.

Other foods swallowed by owls may, however, be digested entirely, such as slugs or earthworms (not an uncommon food item of screech-owls). These leave little or no evidence in pellets, so that an analysis of an owl's diet based entirely on pellets should be viewed with considerable caution. Snowy Owls known to have been feeding on waterfowl produced no pellets with feathers but did produce hair pellets from ingested mice (Brooks 1929). The big owls had simply stripped the meat from the bones of the birds without swallowing feathers.

Fig. 38. Pellets found in woodlands *(clockwise from upper left):* Barn Owl (often oval); Northern Saw-whet Owl (very small); Great Horned Owl (often containing large bone fragments).

Pellet analysis is therefore not as reliable a diet indicator in large owls as it is in medium-sized species, such as Barn and Long-eared owls. Smaller owls present difficulties of their own because of the small size of their prey and the rapid decomposition of their pellets containing insect parts (Marti 1987). Pellets formed after a meal of amphibians disintegrate very rapidly in the absence of binders such as fur or feathers. There may in fact be no pellet at all, but merely a little pile of small bones. The presence of these animals in owl diets may therefore be underreported.

Although pellets in dry areas remain intact for long periods of time, they may be eaten by some arthropods and other creatures. Certain moth larvae, for example, consume the compacted fur. Two very young Desert Tortoises *(Gopherus agassizii)* no more than 6 cm (2.25 in.) long, when placed on a dark tarp with various other objects, made a beeline for four Long-eared Owl pellets some two feet distant and ate substantial parts of these unexpected reservoirs of calcium (P. Gordon, pers. comm. 2007).

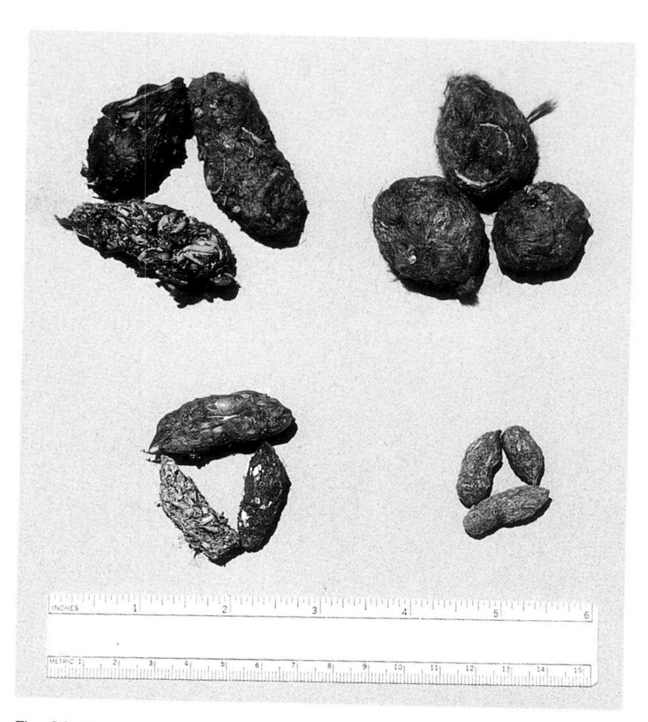

Fig. 39. Pellets found in grasslands *(clockwise from upper left):* Short-eared Owl (note lower pellet almost entirely composed of fragments of Jerusalem Cricket exoskeletons); White-tailed Kite (often pendant-shaped); ground squirrel droppings, sometimes mistaken for owl pellets; Burrowing Owl (Jerusalem Cricket remains, fragments of eggshells [from its own eggs, after hatching]).

AN OWL'S LIFE

THREE ENDEAVORS DOMINATE THE LIFE of an animal—eluding predators, eating, and producing offspring. Although the order of importance is likely the reverse of the above for most animals, owls appear to be extraordinarily tasty to a great many consumers and have evolved sophisticated countermeasures to becoming a meal; escaping an enemy's notice by hiding and by camouflage are common means of survival.

Most owls are expert foragers, although many of their prey have evolved protective behavior and colors as well. The night hunters have countered these with amazingly sophisticated sensory systems that are so effective that owls can seemingly afford wasteful feeding. A crash of rodent populations, however, can profoundly affect these predators, especially their reproduction. Many owls adjust their clutch sizes to the available food supply, and in lean years, they may not breed at all, or the smallest young may become food for the larger ones so that at least some may fledge.

Predators and Predator Avoidance

Predators of Owls

Although adult owls are probably rarely caught by snakes, rattlesnakes *(Crotalus)* are known to feed on nestling Burrowing Owls *(Athene cunicularia)*, as do gopher snakes *(Pituophis)* (Allen 1967), which are also surprisingly competent climbers and collect nestlings in tree holes: one such serpent was found in the process of swallowing a four-week-old Barn Owl *(Tyto alba)* nestling ten feet up in a sycamore *(Platanus)* cavity (P. Bloom, pers. comm. 2006).

Several carnivorans are important predators of owls. Badgers *(Taxidea taxus)* take Burrowing Owls (or their eggs) in their holes, and Coyotes *(Canis latrans)* and foxes *(Vulpes)* can surprise this species as well as the Short-eared Owl *(Asio flammeus)*, another ground dweller. The Bobcat *(Felis rufus)* can fly up near-vertical trees with amazing speed to snag a bird before it fully takes wing. Raccoons *(Procyon lotor)* enter large tree holes and kill owls, and in its montane habitat the slender and agile Pine Marten *(Martes americana)* routinely inspects smaller cavities for delectable contents.

Fig. 40. At a cliff, a female Prairie Falcon *(Falco mexicanus),* enraged by the presence of humans near her eyrie, strikes a Barn Owl that the same visitors, careless rock climbers, have accidentally dislodged from its hiding place.

When the opportunity presents itself, a variety of hawks, eagles, and falcons prey on owls, which, given the chance, return the favor at night. Diurnal raptors are presumably acutely aware of this predation, because most miss no opportunity to dive at and even strike (often to kill) an owl caught by day out in the open. Sometimes they will eat them and sometimes not. In any event, owls have good reason to avoid being noticed by hawks, even when the owl is large and powerful.

The nest of a Cooper's Hawk *(Accipiter cooperii)* contained three mostly plucked bodies of Western Screech-Owls *(Megascops kennicottii)* (Peeters and Peeters 2005), and Red-tailed Hawks *(Buteo jamaicensis)* kill Barred Owls *(Strix varia)* and even Great Horned Owls *(Bubo virginianus)* (M. Carrasco, pers. comm.

2005). A Northern Goshawk *(Accipiter gentilis)* caught a Great Gray Owl *(S. nebulosa)* in midair (Duncan 2003). The ground under a fence post much favored by a Golden Eagle *(Aquila chrysaetos)* was littered with the feathers of Burrowing Owls, a species that, because of its open-country habitat, is a frequent victim of diurnal raptors.

As if that were not enough, owls fall prey to larger owls. The Barred Owl includes screech-owls *(Megascops)* and Long-eared Owls *(Asio otus)* in its diet (Holt et al. 1999). Both the Long-eared Owl and Short-eared Owl have been recorded as prey of the Great Horned Owl (Bent 1961), which is also known to take Barn Owls (Rudolph 1978). The pellet of one Great Horned Owl contained the nearly intact foot of a Burrowing Owl (Peeters 1963). Singing Flammulated Owls *(Otus flammeolus)* at once fell silent and froze upon hearing the hoot of a Spotted Owl *(Strix occidentalis)* (Marshall 1939), although other observers reported males reacting vocally to Spotted Owl recordings (Marcot and Hill 1980), and other small owls commonly respond to Spotted or Great Horned Owl playbacks (P. Bloom, pers. comm. 2006). Nevertheless, a captive Western Screech-Owl kept indoors instantly stopped singing and dove into its roost box upon hearing a Great Horned Owl hooting from the chimney of the house. Some owls, including the Burrowing Owl, have been also known to eat their own kind (Robinson 1954).

Going Undetected

It is a measure of the persuasiveness of owls' disguises that these birds exist at all, given their popularity as food. The somber colors of their feathers are arranged in subtle patterns so that their owners resemble bark (often with lichen), dead limbs, branch patterns, dead leaves, or rocks or stony soil. These masquerades are supported by appropriate behaviors, foremost remaining motionless. If they need to move their heads, they usually do so very slowly, as is typical of a number of Neotropical rainforest birds that also seek to escape detection by predators or prey. In addition, small owls press themselves against tree trunks, literally becoming bumps on a log, or partly insert themselves into little tree cavities and crevices, and Short-eared Owls seem to pretend they are tussocks in their sere grassland roosts. Some species tighten their feathers to make themselves slim and, standing

Fig. 41. Even a very large owl can blend in so well with a dead tree that it virtually disappears in plain view. Great Horned Owl shown.

bolt-upright, raise their ear tufts (if present) to turn themselves into convincing broken limbs complete with smaller branch stubs on top.

When they find themselves discovered, many owls may rotate their bodies a bit, compress their feathers, and draw one folded wing forward partway across the breast and belly to cover these paler undersides. The species that do so have vertical, prominent rows of light and dark spots and markings running down the scapulars and the leading edges of the wings that look like bark ridges and fissures. Simultaneously, many fold up their facial disks so that they look prunefaced, eliminating thereby the eye-catching circle of the disk, and they nearly close their eyes, watching the observer through the merest slits between their eyelids. In this way, they hope to escape detection. If eye contact is made, however, they often fly away.

For small and midsized owls, tree cavities would offer ideal hiding sites (and day roosts), with a fairly constant microclimate, safe from the eyes of mobbers and most predators, were it not for one major problem: they offer no chance of escape should a predator come through the front (and usually only) door. Small

Fig. 42. Typical concealing posture of a small owl, seen here in an Elf Owl.

owls can choose holes tight enough to exclude some predators, and Western Screech-Owls, which generally live in areas where there are no Pine Martens, routinely do use tree cavities (and nest boxes) for roosting. However, only the females of Northern Saw-whet Owls *(Aegolius acadicus)* and Flammulated Owls—species that may share their habitat with martens—take advantage of such holes, and only during the nesting season. Scratching on a tree trunk below the hole instantly causes the occupant to peer out, although screech-owls generally cannot be bothered (Marks Cannings, and Mikkola 1999). A tree trimmer ascending a sycamore provoked a nesting Northern Pygmy-Owl *(Glaucidium gnoma)* to stick her head out of the nest hole while the climber was still far below. Disturbing nesting owls in this fashion on purpose, however, is to be discouraged.

Fig. 43. Often, Spotted Owls require no more than the forest canopy overhead to perch comfortably.

Some owls seasonally use tree hollows because of their favorable microclimate but switch to more open roosts in summer. The larger owls, from Spotted Owls on up in size, may perch quite exposed, high up in the trees.

Most owls close or nearly close their eyes as they perch during the day, probably as much for concealment as for sleeping. Many birds seem to know that large, yellow eyes belong to something dangerous, such as an owl (even when there is no owl!) and quickly come together to harass the predator. However, they may also discover and mob an owl that has its eyes closed, and dark-eyed owls do not get a free pass, either.

Mobbing

Mobbing is the vehement, usually noisy harassment of a predator by potential (though at times unlikely) prey birds of one or several species. For example, it is not uncommon for curious hikers who follow a racket in the woods to discover a mob composed of Western Scrub-Jays *(Aphelocoma californica)* or Steller's Jays *(Cyanocitta stelleri)*, or both, berating a Western Screech-Owl or a Great Horned Owl. Mobbing is widespread; even Northern Pintail ducks *(Anas acuta)* have been observed to mob a Short-eared Owl (Erickson 1955), and several Elf Owls *(Micrathene whitneyi)* mobbed a Great Horned Owl (Boal et al. 1997). The imitation of a Great Horned Owl song in November brought in three Western Screech-Owls reviling the author in his central California woodland yard.

Fig. 44. Although the two black blotches on the back of a pygmy-owl's head often do not suggest a face, sometimes they do look like eyes. Juvenile Northern Pygmy-Owl shown.

There is evidence that predator recognition and mobbing are learned, at least in part and in some bird species. Juvenile and adult Gray Jays *(Perisoreus canadensis)* were fed experimentally in the presence of a stuffed Great Horned Owl. The juveniles were largely indifferent to the presence of the decoy but became more hesitant when they observed the agitated behavior of the adults, which, however, gradually habituated to the owl (Montevecchi and Maccarone 1987).

A study of mobbing behavior using visual and auditory stimuli of the Eastern Screech-Owl *(Megascops asio)* revealed that a stuffed owl alone was a weak attractant, whereas the taped song and the combination of decoy and tape were much more effective in drawing in mobbers, with the combination initiating mobbing behavior that was much more intense and longer lasting than the response to the recording alone. Obviously, the taped vocalization greatly increased the chances of the owl being discovered (Chandler and Rose 1988).

Other experiments involving the same owl species showed that the attraction to the "vocalizing" decoy was strongest when the tape played the spring and early summer advertisement song. By contrast, the territorial song, which is given later in the year, elicited mobbing from far fewer species and fewer individuals. The most frequent mobbers were members of species most commonly eaten and were dominant males of permanent residents. If the owl was near its nest site, the songbirds mobbed more

picked the bird like a fruit, brought it inside, and placed it in his Christmas tree, where it sat calmly for a couple of hours before being returned to its outdoor perch (E. W. Jameson Jr., pers. comm. 2005).

Other owls, handled on their nests, seemingly feign death, at times to the point of rigor mortis. Incubating female Barn Owls become very lethargic, appearing nearly comatose, and allow themselves to be turned onto their backs while the eggs are examined (S. Simmons, pers. comm. 2005), and Western Screech-Owls may react similarly. Snowy Owl young stiffen their bodies and extremities when picked up or placed on their backs (Watson 1957), and others became so rigid that they could be balanced on a pointed rock (Parmelee 1972).

Hunting and Eating

A California savanna of grassland and scattered oaks may be home to both a pair of Red-tailed Hawks and a pair of Great Horned Owls, species that feed on similar-sized prey. They do not compete with one another for the same food resources because one species is a day hunter and the other is active at night, and there are differences in prey preference as well. They may successfully nest within about 100 m (300 ft) of each other and, on at least one occasion, have nested in the same tree, the hawk in a stick nest near the top, the owl 12 m (40 ft) below in a cavity (P. Bloom, pers. comm. 2006).

Contrary to popular opinion, not all owls hunt at night. Some forage during the day (for example, the pygmy-owls) or during the day as well as at night (such as the Snowy Owl and the Short-eared Owl), thereby reducing competition with other owls. Their numbers are augmented when there are young to feed, a period when even Barn Owls, normally highly nocturnal, may be abroad in daylight. There are also crepuscular hunters, that is, species that hunt chiefly just before dawn and at dusk (when the majority of owls are most active), and others that are active throughout the night. Barn Owls have no problem foraging in complete darkness, hunting by ear. Perversely, Burrowing Owls, which loiter so visibly about their burrow entrances during the day, are in fact chiefly (though not entirely) crepuscular and nocturnal hunters.

Fig. 48. When there are young to raise and food is scarce, Barn Owls may forage in the daytime.

The majority of North American owls forage most often in one of two ways: by still- or perch-hunting and by search flights. Of the two, still-hunting is the most common; the owl perches on a branch or similar lookout to wait for prey and then drops on or flies short distances to it. Periodically (usually within a few minutes), it moves on to a new perch. Because still-hunters have relatively short wings and therefore tend to have relatively heavy wing-loading, prolonged perching saves much energy. Spotted Owls and Western Screech-Owls hunt this way, for example.

By contrast, owls that quarter over open ground have longer wings and lower wing-loading and can therefore stay aloft with less energy consumption. Flying much like the diurnal Northern Harrier *(Circus cyaneus)*, at a height of a few feet and at a relatively slow speed, they cover great areas of suitable foraging grounds and, having located prey, may hover briefly with dangling legs before dropping on it. The Short-eared and Long-eared owls are examples of such active searchers, and the Barn Owl uses both methods equally.

Direct pursuit is relatively unusual among owls but not unknown. A Long-eared Owl caught moths with its beak on the wing, hovering intermittently and flying about (Sleep and Barrett 2004), and Eastern Screech-Owls also catch flying insects (Sutton 1929), as do Burrowing Owls. Some owls pursue flying bats, and the formidable Snowy Owl catches birds as large as ducks and even geese in the air (Parmelee 1992). Great Horned Owls trained

Fig. 49. Entrance to a Burrowing Owl's hole. The adults have brought assorted junk—a strip of carpeting, a jackrabbit's foot, a corn cob—perhaps for their young to practice prey capture. A mouse-half awaits someone with an appetite.

in falconry chase rabbits (a common prey item found in their nests). Northern Pygmy-Owls actively hunt birds and reptiles, and Burrowing Owls may run or hop after prey (usually arthropods) and hover like American Kestrels, sometimes at altitudes of well over 30 m (100 ft) (Sprunt 1943). Great Horned Owls walk through grass, around bushes, and in other promising areas in search of small mammals and invertebrates (Houston et al. 1998), and juveniles also were observed to hover, suggesting that this method of foraging may be more widespread in owls than generally realized (Smith and Smith 1972). Northern Saw-whets hover well, as do Long-ears.

In Florida, experiments with local Burrowing Owls showed that these birds placed cow dung around their burrows for the express purpose of attracting dung beetles, which the owls use as food. Such baiting is a form of tool use that mirrors the use of bread, feathers, and insects by some herons to attract fish (Levey et al. 2004). Another study found cow dung scattered inside the nest burrow and tunnel and around the entrance; systematic trapping showed that such enhanced burrows attracted more food arthropods than undecorated burrows (Smith and Conway 2007). The owls' dung gathering may account for this species' puzzling habit of bringing all sorts of junk to its nest burrow's entrance, perhaps mistaking it for feces (although strongly patterned objects are often disproportionately represented). One pair, nesting in Fremont (Alameda County), actually retrieved dog droppings from a nearby dog exercise area, besides other odd items (S. Ferreira, pers. comm. 2005). It is interesting to speculate how the owls determine a given object is in fact mammalian waste, the smell of which has been thought by some to mask the odor of the birds to predators. In any event, these collectibles eventually come in handy, providing the young with objects upon which they can practice their pouncing, footing, and biting skills.

Although some owls specialize on small and usually plentiful prey, others, both young and old, having decided that a given animal is edible, are real brawlers and do not mind a rough-and-tumble struggle with even very large prey, which may involve rolling about on the ground with the victim. All the while, they knead it with powerful feet, in an effort to penetrate a vital organ, and bite at the neck and head. A study of five wild-caught Eastern Screech-Owls revealed, however, that, given the choice, they more often opted for the smaller prey (Marti and Hogue 1979). Oddly, a captive young Northern Saw-whet, loose in a large room but perhaps not very hungry, watched live prey (moths, crickets, and

Fig. 50. A captive juvenile Western Screech-Owl subdues a shoe.

mice) with great interest but would generally go after them in earnest when they were about to disappear into cover.

Their hunting skills notwithstanding, owls can starve to death when rodent populations crash, and severe winters are deadly for various species, even the snow-adapted Great Gray Owl. However, owls can fast for a very long time, even the small species. A captive Western Screech-Owl went without food for a week, and a caged Northern Pygmy-Owl, kept indoors, unaccountably refused all offered mice (its usual diet) for 22 days, during which it drank, vocalized, and produced droppings, though sparingly. Prior to the fast, the bird had been given a surplus of mice; breaking the fast, the owl ate one entire mouse and the head of another, and then produced a monstrous pellet measuring 45 mm (1.75 in.) long and 10 mm (.4 in.) thick, much larger than the peanut-sized pellet typical for this tiny owl (E. Hobbs, pers. comm. 2007).

Feeding

If the prey is small, the owl clutches it, crouches over it, and, with its eyes closed, feels usually for the head with its rictal bristles to administer killing bites to the back of the head, crushing the cranium. Having killed it, an owl may momentarily remain crouched on the ground with spread wings and tail and peer about. It may eat its prey then and there or carry it to a perch before consuming it.

Many owls are inclined to swallow prey whole if possible, and some can manage amazingly large items; watching a good-sized Botta Pocket Gopher *(Thomomys bottae)* pass down the gullet of a Barn Owl in one piece is a sight not soon forgotten. Different from hawks, owls maneuver large items into the wide-open gape with odd, vigorous down-and-sideways sweeps of the head, which is then raised high to ingest the food with up-and-down pumping of the head, as a hawk does. Prey that is eaten whole usually slides down headfirst (although snakes may go down tail first, too). The animal may also be decapitated, the owl eating nothing but the head, or eating the head along with a leg or two and saving the rest of the body for later.

Sometimes, the digestive tract of the quarry is removed. One lucky Burrowing Owl in Saskatchewan unwittingly saved its life by discarding a dead strychnine-poisoned ground squirrel's guts before feeding on it, thereby avoiding the greatest amount of the toxin ingested by the rodent (James et al. 1990). Pygmy-owls, as well as most large owls, pluck avian prey before eating it in large chunks. Short-eared Owls, however, do not pluck birds they have killed, swallowing small ones whole after taking off the wings (Holt and Leasure 1993).

Prey, especially very large items, may be eaten in installments, with the owl in effect feeding on carrion. For a period of nearly a week, one Great Horned Owl dined at intervals on a domestic rooster it had killed; because it was in the middle of a Canadian winter, the owl had to intermittently sit on the dead fowl, putting it into the microwave, as it were, to thaw it sufficiently for consumption (Duncan 2003). Northern Saw-whet Owls, too, will lie on frozen prey previously stored on branches (Cannings 1993), as do Eastern Screech-Owls (Duncan 2003); it is likely that this behavior is common to all owls living in snowy regions.

Food storage (caching) is a useful ploy for a raptor that lacks a crop in which to carry another meal or when a food supply is temporary or unpredictable, or when much food needs to be on hand to feed young. At an Indiana farm, an Eastern Screech-Owl cached 24 one-day-old chicks in six hours, apparently for its personal use, as breeding season had not yet commenced (Cope and Barber 1978); another, in Canada, stored voles *(Microtus)* on a rafter inside an old shack, replenishing its stash seemingly daily (Phelan 1977). In central California, Northern Saw-whet Owls cached live mice, which they had paralyzed, in their nest boxes to feed their young (G. Monk and S. Lynch, pers. comm. 2004). Some owls cover cached food with plant debris or snow.

The amount of food consumed daily by an owl is related to its size, with the smaller species requiring disproportionately more fuel per body mass. Their greater surface area relative to their body volume allows essential heat to escape more quickly than it would from a large-bodied owl (a cake cools off slowly, a slice very quickly). Mean consumption of a Great Horned Owl was 4.7 percent of its body weight per day, and of a Burrowing Owl, 15.9 percent (Marti 1973). The tiny Northern Pygmy-Owl requires a daily food intake of about 50 percent of its body weight (Johnsgard 2002).

Because it is often easy to find dead rodents or their parts under roosts, owls would appear to be profligate (or at least careless) feeders, especially compared to diurnal raptors, which jeal-

Fig. 51. Owls, presumably accidentally, drop lots of prey and, in the absence of movement, cannot find them (here, a Dusky-footed Woodrat [*Neotoma fuscipes*]).

ously guard food and defend it against other predators; a meat-eater's meal is after all not easy to secure and requires a sizeable outlay of energy. It is, however, likely that owls simply cannot see food dropped accidentally in poor light and cannot retrieve dead prey, which of course makes no sound.

In any event, so much prey is lost, especially during the breeding season, that mammalian predators such as Gray Foxes *(Urocyon cinereargenteus)* and Coyotes and probably other carnivorans learn to regularly patrol under owl roosts and nest sites, as evidenced by their droppings. It is likely the abundance of dead wood rats *(Neotoma)* and half-eaten mice dropped by owls that attracts Turkey Vultures to woodlands. In much of California and elsewhere in the West, these scavengers cruise for hours over wooded areas and have an uncanny knack for locating even small morsels in the leaf litter under the trees. The hunting prowess of owls must play a major role in cycling rodent biomass through woodland ecosystems, a service that is probably underestimated.

Owl Foods

Although owls seem to be consummate rodent specialists, the various kinds in fact feed on an astonishing variety of foods. For example, several species on other continents prey almost exclu-

Fig. 52. An owl's head can look small while it preens its neck. This Northern Pygmy-Owl obviously eats vertebrate prey: note the blood on and around the beak.

sively on fish and have nearly naked legs and feet. Others catch dragonflies in flight—no mean feat—and pounce on crabs, and many feed on a virtual smorgasbord, from insects to foxes.

North American owls, depending on the species, may prey on scorpions, moths, fish, reptiles, mice, rats, rabbits and hares, birds large and small, and even skunks and raccoons. Two young Bald Eagles *(Haliaeetus leucocephalus)*, one a chick and the other a fledged juvenile, were found decapitated and partly eaten, strong evidence that they fell victim to Great Horned Owls (Hunt et al. 1992). There is also the constant danger that the smaller species of owl may become a meal for its larger cousins.

With their alarm calls, Black-capped Chickadees *(Poecile atricapillus)* can indicate whether they have spotted a Great Horned Owl or a pygmy-owl (Templeton et al. 2005), demonstrating that they, along with other small birds, have little to fear from the big predator but are not infrequent prey of the dwarf.

Owl Ecology

No matter what they feed on, owls are generally at or near the end of food chains (only when they are eaten by another predator do they form a penultimate link) and at the top of consumer pyramids. Acres of grassland or forest are required to feed the quantities of voles and mice to sustain most owls and their young.

Within their communities, the various owl species appear generally extremely well adapted to their ecological niches. Not only are their modifications for life in the dark highly refined, but they are also competent predators of mostly elusive prey. It is remarkable that the majority of species live in wooded habitats where navigating safely at night is the most difficult; the abundance of rodents active at night in forests and woodlands doubtless played a major role in the evolution of owls.

Owls are frequently called the nocturnal counterparts of the hawks, eagles, and falcons that hunt by day, and although they occupy similar ecologic niches, there are differences beyond the obvious anatomical and temporal ones. A study comparing the diets of the Red-tailed Hawk and its apparent night counterpart, the Great Horned Owl, found that although their use of prey species overlapped 50 percent, the hawk's dietary diversity was greater than that of the owl, which ate significantly more invertebrates and killed smaller vertebrate prey. Both raptors are generalist predators, but they feed largely on different prey species in the same locality (Marti and Kochert 1995). Many owls seem to forage over much smaller areas than their presumed diurnal counterparts, harvesting more biomass per unit of land.

Beginning owlers are often surprised to learn that a not-so-large forest can hold several species of owls. A walk at dusk or dawn through an oak-laurel woodland at the right time of year in central Alameda County, for example, may reveal that it can provide habitat for five and sometimes even six species of owls, which a persistent owl student can count on finding sooner or later. Each species has certain preferences for nest site, foraging time and area, and prey, although there are some overlaps. The Barn Owl chiefly keeps to more open, grassy areas and grassland "bays" along the edges. The Western Screech-Owl uses the riparian trees (though not exclusively), as does the Northern Pygmy-Owl, which however forages at different times. The Northern Saw-whet Owl haunts the densest parts of the woodland, and the Great Horned Owl utilizes all parts. The sixth species that can be found here is the Long-eared Owl, which is an irregular breeder in this habitat, utilizing riparian corridors for nesting, while hunting in more open settings.

Apart from the diet, which may be very similar, other aspects

Fig. 53. This north-central coast oak-conifer woodland and coniferous forest, like some other habitats, can hold as many as six owl species.

of their ecological niches separate these species sufficiently to allow them to live sympatrically. Western Screech-Owls and Northern Saw-whets, for instance, both inhabiting an Idaho Douglas-fir forest interspersed with mountain shrub, bunch grass, and some riparian zones, chose different roost sites, with the Northern Saw-whet hiding in the most concealing dense foliage at the end of a branch, and the screech-owl roosting next to the tree trunk where its coloring merged with that of the bark (Hayward and Garton 1984).

Spotted Owls and Great Horned Owls in a pine-oak forest in northern Arizona showed substantial overlap in diet composition, as well as complete overlap in the size range of prey, which included 63 percent (for the Spotted) and 62 percent (for the Great Horned) mammals, comprising 94 and 95 percent of prey biomass, respectively. However, Great Horned Owls took more quarry active by day than did the Spotted Owls, and although they could compete for the same foods that might be scarce in some years, the two owl species may minimize competition by chiefly foraging in different habitats and at different times when different prey animals are vulnerable (Ganey and Block 2005). Where a complete or near-complete dietary overlap between owl species does occur, it is in areas and years of extreme abundance of certain rodents.

In parts of Mexico, the ranges of the Barred Owl and the Spotted Owl, two species usually fairly similar in size, have historically overlapped; there, the local race of the Barred *(Strix varia sartorii)* is large and that of the Spotted *(S. occidentalis lucida)* is small, so that the two owls apparently occupy different niches with minimal competition. A diet study of sympatric Barred and Spotted owls in Washington State, where the two owls (*S. v. varia* and *S. o. caurina*) are much closer in size, showed that the Spotted caught more arboreal and semiarboreal mammals than the Barred, which also took more diurnal prey as well as animals found in wet habitats (Hamer et al. 2001). As the Barred Owls advance in California, they displace the Spotted Owls, as they have done in Washington and Oregon, pushing them to the margins of their former territories, where they cannot breed and do not even sing (B. Woodbridge, pers. comm. 2006).

Sympatric existence between owls, however, obviously does not always result in conflict. In Colorado, for instance, Barn Owls sequentially reused Great Horned Owl nest sites within the same years (Andersen 1996), taking advantage of the same resource by temporal separation. In southern California, Barn Owls in one case shared the same nest tree with Great Horned Owls (with their cavities on opposite sides of the trunk) and in another with Western Screech-Owls, with the larger species' nest hole only two feet above that of the smaller one (P. Bloom pers. comm. 2006).

A study of four sympatric owls (Great Horned, Long-eared, Burrowing, and Barn) in north-central Colorado showed that although there was much overlap in the prey they ate, each species specialized on different groups and sizes (and hence weight) of prey. The Great Horned Owl harvested not only the largest prey by far but also selected from the greatest size range, whereas the Long-eared Owl was the most restricted in its selection of prey species, choosing mammals almost entirely and especially seeking out voles. Overall, prey size was likely the chief determinant in feeding niche segregation (Marti 1974). Northern Saw-whets in southwestern Idaho selected prey from a much more limited size range than did the sympatric Western Screech-Owls (Rains 1997).

Along with the Long-eared Owl, the Short-eared Owl, too, relies heavily on small mammals, mainly voles, whose populations are cyclic; in the San Francisco Bay Area, for example, the population of California Meadow Voles *(Microtus californicus)* peaks every three or four years (Ingles 1965). Owls, responding to such changes, can show dramatic differences in the number of adults breeding, the numbers of eggs laid, and the numbers of fledglings. Fat, well-fed female owls lay many eggs (Marks, Cannings, and Mikkola 1999). The management of grasslands for waterfowl in Solano County by the California Department of Fish and Game in 1986 and 1987 appeared to greatly increase rodent populations, which in turn may explain a dramatic increase in nesting Short-eared Owls (Larsen 1987). For unknown reasons, however, breeding Bay Area Short-eared Owls have greatly decreased in numbers, healthy vole populations notwithstanding.

The Snowshoe Hare *(Lepus americanus)* is well known for its population cycles: hare numbers peak at regular intervals of a bit less than 10 years and then suddenly decline. In a Canadian subarctic forest, during the increase phase of the cycle, nearly all resident Great Horned Owls bred and raised large broods, with high survival rates of juveniles and a near doubling of territory-holding owls. As the hare population declined, postfledging mortality of young owls was high, and floaters (non-territory-holding owls) emigrated or died before territorial birds did (Rohner 1995).

The productivity of a southern Texas Barn Owl population was greatly reduced when the overall small mammal population decreased in its availability as prey, although the changes occurring in a single small mammal species had no immediate effect on the owls' numbers or breeding success. An abundant blackbird population, however, could not sustain successful owl breeding, showing that both quality and quantity of available food are important (Otteni et al. 1972).

In northern Europe, Great Gray Owls feed on four to six species of voles during the nesting season, whereas in California, only one (and marginally two) are available, although pocket gophers *(Thomomys)*, too, form part of the diet but appear to be a maintenance food, insufficient in numbers to permit breeding. Because this owl does not readily switch to other types of prey, it may not breed in California regularly when vole populations are low (Winter 1985). Gophers may never be numerous enough to sustain a family because their territoriality limits their populations, but voles are nonterritorial and can saturate a habitat.

Owls, as carnivores, occupy the uppermost trophic levels in their ecosystems, and as such, their populations are by necessity limited. Some appear to be K-strategists, meaning they are animals that, living in a stable environment, are long lived, mature slowly, and can survive as species by producing small clutches or litters; their populations are held in check by the carrying capac-

Fig. 54. A fledgling Barn Owl, as an r-strategist, may not live for many years.

ity (K) of their environment. The Northern Spotted Owl is an excellent example of a K-strategist, and its population size is density dependent.

If K-strategists appear almost environmentally conscious (adjusting their numbers to what the land can sustain), their counterparts, the r-strategists, are the uncaring proletariat: r-strategists breed young and often; they have large broods and, being density independent, have a high intrinsic rate of population increase (*r*), as well as high mortality, dying at an early age. They do well in unstable environments and can take advantage of environmental windfalls, such as novel habitats and new foods. The Barn Owl is in many ways a typical r-strategist; it breeds and dies young, has many offspring, and has the ability to find new resources (such as agricultural fields), sometimes at great distances, and exploit them (Marti 1999). In short, as a species, this owl basically has the strategy of a rat.

Most animals, including owls, are neither pure K-strategists nor pure r-strategists, but fall somewhere in between, at times leaning more in one direction or the other. Both Great Horned Owls (apparent K-strategists) and Barn Owls in tandem increased or decreased their productivity in southern California in response to prey availability (P. Bloom, pers. comm. 2006). The larger clutch response of Great Horned, Long-eared, and Short-eared owls in the presence of a greatly increased food supply (cited above), for example, smells suspiciously of r-strategy.

Although most owls that might experience a cold and snowy winter have sufficient fat reserves to survive a temporary inaccessibility of prey, have dense feathering to keep them warm, and can fast for extended periods, the Barn Owl is limited in its distribution by its lack of adaptations for cold weather. Many die during unseasonable cold snaps, or abandon their eggs (Smith and Marti 1976). In Northern Utah, at the northern limit of its range, 77 individuals were found dead in the winter of 1981/82, all of them severely emaciated, having died from the extreme cold and unavailability of voles because of the deep snow cover. There followed a 40 percent decline in breeding attempts in the following spring (1982), and only 1.5 young fledged per nesting attempt (as opposed to 4.1 young in the four previous years), but 1983 proved a banner year because of an influx of individuals (recognized as immigrants by their lack of legbands) and nesting by adults that had failed to do so the previous year (Marti and Wagner 1985).

The fat (lipid) reserves of a European race of the Barn Owl (which was long believed to have low fat reserves) were compared with those of the sympatric Tawny Owl *(Strix aluco)* and the Long-eared Owl, both of which do not ordinarily suffer high winter mortality. Because the adiposity of the three species was similar, it was concluded that the Barn Owl's higher mortality in winter was due to poorer feather insulation, diet, and its hunting behavior (Massemin and Handrich 1997). Surprisingly, the occasional Barn Owl can survive for a long time: one wild individual reached 34 years of age (Keran 1981), r-strategy nothwithstanding.

Reproduction

Territoriality

The defense of an area of whatever size and for whatever purpose by animals against conspecifics is known as territoriality. Most often, this activity serves to protect a food-rich area or an area used for reproduction or both. Like other raptors, owls have home ranges that provide their food, and their breeding territories may equal these areas or may not be included in them. Some owls, such as Short-eared Owls, are semicolonial and may breed in close proximity to one another (and thus defend very small areas around the actual nest only) but forage over much greater areas. The home ranges of five radio-tagged female Great Horned Owls averaged smaller than those of nine males, perhaps because of the females' nesting activities, and three territorial male neighbors showed no home range overlap. All-night observations revealed that these owls typically used only about a third of their home ranges, and sometimes considerably less (Bennett and Bloom 2005).

Because prey animals, especially vertebrates, are rarely concentrated in a given area, the territories or home ranges of owls are often sizable, to supply the raptors with sufficient food, although some species, such as the Western Screech-Owl, probably because of its diverse diet, can have surprisingly small territories. The requirement for ample food becomes especially critical when there are young to raise. Territorial defense, during which talons may be liberally used, is normally directed at members of an owl's own species that might otherwise seek to usurp the same resources.

Excluding your own kind is known as intraspecific territoriality, often observed in spring when a male American Robin incessantly attacks its own reflection in a window, perceiving it as a conspecific intruder. Other species, however, may sometimes be attacked as well when they seek the same resources; this is interspecific territoriality, as seen in a hummingbird chasing off hummers not only of its own kind but other species as well in its defense of "its" feeder. Owls, like songbirds, announce the presence of an occupied territory by vocalizing, thereby warning off trespassers and avoiding armed conflict.

Some owls defend their territories or home ranges year-round, the Western Screech-Owl being an example. If food is plentiful, territories tend to be small, and large if prey is widely dispersed. The breeding territories of Short-eared Owls increased in size when the chief prey, meadow voles, decreased in numbers (Clark 1975). On the other hand, 14 territories of Western Screech-Owls fit into a mere 6.4 km (4 mi) of riparian woodland in southern California, indicating abundant prey and nest sites (Feusier 1989). All screech-owls as well as the Elf Owl can be poly-territorial, defending areas around several potential nest cavities (Gehlbach and Gehlbach 2000).

Pair Formation and Courtship

Although a mated pair of owls, if nonmigratory, may remain within its territory year-round, the two pair members do not necessarily roost together. Some Barn Owl pairs, for example, wait until breeding commences, whereas others seek out each other's company as early as November (Marti 1992). Great Horned Owls generally keep their distance from each other until close to egg-laying.

As the breeding season approaches, male owls vocalize with increasing frequency, both to announce their territories and to attract females or, if already present, to reestablish the pair-bond. Because Great Horned Owls are early breeders and may be on eggs in January, the deep, booming hoots of the males echo through the woods as early as October; this species may in fact be heard throughout the year.

Fig. 55. How a male Barn Owl attempts to impress a prospective mate. (Photo by Steve Simmons.)

Even Western Screech-Owls, which nest in spring, become vocal in fall, after a period of relative quiet as the young of the previous breeding fledge and disperse. The males of the migratory Flammulated Owl sing to attract females, which, in turn, if they have arrived on their breeding grounds later, give food solicitation calls to attract a suitor. A male Barn Owl may pile up extravagant numbers of mice in a nest box to lure a partner.

Studies of the Eurasian Tawny Owl, a close relative of the Barred Owl, showed that the songs of the males conveyed information about the number and intensity of parasitic infections (Appleby and Redpath 1997), thereby supplying details about the singers' health status and, by extension, fitness as food providers.

Once a male owl has attracted a female's attention, he leads her to one or more prospective nest sites, frequently making food offerings to coax her. Barn Owl males fly in and out of cavities while calling, and a Western Screech-Owl suitor carries food into cavities, hoping the female will follow (Cannings and Angell 2001). Ultimately, it may be the female that selects the nest site.

Except for the Burrowing Owl (in which the male is larger), the females of North American owls are bigger than males (up to 40 percent), a condition that also occurs in diurnal raptors. A great many hypotheses have tried to explain this phenomenon, called reversed sexual dimorphism, some of them based on the observation that it is most commonly observed in raptors that feed on vertebrate prey. The smaller mate feeds on smaller prey than his larger partner, which would lower intersexual food competition. In support of this hypothesis, it has been pointed out that most insect-eating owls show little or no sexual dimorphism because the prey is plentiful, whereas great sexual dimorphism is found in the chiefly vertebrate-eating Great Horned Owl, whose prey, though encompassing a much bigger size range, is also much more thinly distributed (Earhart and Johnson 1970). However, the most extreme example of reversed sexual dimorphism is the Mottled Owl *(Ciccaba virgata)*, a tropical species that is highly (though not predominantly) insectivorous. Perhaps the female's larger size makes it easier for her to bully the male out of food to supply her and her young, or it facilitates incubating and brooding the often numerous eggs and young, tasks that, in owls, are generally left exclusively to the female. It is usually she, too, who must deal with predators threatening the nest, although both sexes may participate in this defense, which can be fierce. One

Fig. 56. Juvenile of the neotropical Mottled Owl.

hypothesis postulates that inasmuch as colonial species mount a communal defense against predators, larger females are not necessary; the Burrowing Owl fits this model (Mueller 1986).

Whatever the reasons, male owls are fervent in their courtship and endearing in their endeavor to impress prospective partners. They shower them with gifts, usually in the form of numerous dead rodents. They sing or may engage in fanciful display flights, as does the Long-eared Owl, which zigzags between trees, flies with exaggerated deep wingbeats (a sort of aerial strutting), and claps his wings. Slapping the wings together under the body may, however, be directed as a warning at other males, and females sometimes wing-clap also. The male Short-eared Owl not only sky-dances, performing aerial swoops, wing-claps, and acrobatics, grapples and pinwheels with rivals in midair, and dives by day and night but also delivers a mouse with a little swaying dance prior to copulation, and the Great Gray Owl, in a fine show of machismo, repeatedly plunges into snow as if in pursuit of prey (Nero 1980). Barn Owl males rise in the air and spiral down, clapping their wings, and chase after females in a dodging flight, again with wing-claps.

Once a pair has come together, the bond uniting them is strengthened by allopreening, where pair members take turns nibbling each other about the head. Owls are a faithful lot, and pair-bonds are often permanent, though there are exceptions (especially among highly migratory owls), and the various species are usually monogamous, with polygyny (a male breeding with more than one female) uncommon but not unheard of (Marti 1990). Also, DNA studies of a few species have revealed offspring resulting from extra-pair copulations, the briefest of flings by females with intruding males.

Actual copulation lasts but a few seconds, nearly an anticlimax after all the efforts leading up to it. However, frequency of mating seems to compensate for this brevity, and the act of copulation in itself appears to contribute to pair-bonding. Barn Owls copulate hundreds of times in the two weeks prior to egg laying, indeed before the female is even fertile, and mating continues after the young have hatched (Duncan 2003).

Nests and Eggs

Some of the larger owl species often utilize the old stick nests of diurnal raptors and other large birds, and two, the Short-eared and Snowy owls, the only owls to construct any kind of nest at all, build rather rudimentary ones on the ground. The rest are chiefly cavity nesters, with the Barn and Burrowing owls sometimes excavating their own nest chambers in dirt. Most, however, use preexisting tree holes (sometimes as an alternative to stick nests) or rodent burrows or other cavities and niches, natural or human-made, including next boxes. The female Elf Owl may start spending most of her time in the nest hole a week or more before laying eggs, perhaps to keep competitors from using it (Henry and Gehlbach 1999).

Owls' eggs are usually much rounder than those of chickens, probably because a more spherical shape takes up less room per volume in confined quarters such as tree holes. Owls that use open platform nests have more conventionally shaped eggs (Johnsgard 2002), although, perversely, those of the Great Horned Owl, a confirmed platform user, are said to be the closest to spherical of all birds (Houston et al. 1998). Barn Owls refuse to conform,

Fig. 57. Most owl eggs are more spherical than other birds' eggs *(clockwise from upper left):* Elf Owl egg, two types of Barn Owl eggs, Great Horned Owl egg.

laying both types of eggs, spheroid and ovoid. Because owl eggs are usually well hidden and well defended, they lack protective coloration and are plain white (unless discolored by nest debris).

The contents of owl eggs, too, differ from those of chicken eggs. The whites of an infertile egg of a captive Western Screech-Owl, when fried sunny-side up, were bluish in color; the taste, probably chiefly provided by the fats in the yolk, was very reminiscent of the odor of the House Mouse *(Mus musculus)*. Oddly, the owl was fed mainly a diet of chicken hearts and, occasionally, deer mice *(Peromyscus)*, not House Mice.

Owl eggs are often very large relative to the size of the bird and represent a considerable investment of energy, especially in some of the smaller species, which may produce big clutches (compensating for high mortality). The young of the Flammulated Owl, a highly migratory species, benefit from quickly achieving maximum size before the fall migration. Prior to laying her extra-large eggs, the female Flammulated stops hunting on her own and basically sits around all night eating the arthropod bonbons delivered by her swain, gaining up to 68 percent in weight (mass) (McCallum 1994a).

Female owls in general during breeding season can weigh half again as much as at other times of the year (Marks, Cannings, and Mikkola 1999). A Great Gray Owl may starve herself as she puts all her energy into her eggs. Only the females (with the possible exception of the Short-eared Owl) incubate the eggs and brood the young upon hatching; without fat reserves, they would have to leave their eggs or young to forage should the weather prevent their mates from provisioning them adequately.

Clutch sizes vary greatly not only between the different owls but at times also within a species in response to prey density. Snowy Owls may lay as many as 11 eggs, fewer when food is scarce, and Short-eared Owls as many as 10, again when voles abound. By contrast, Great Horned Owls usually lay two eggs, occasionally three. In peak prey years, however, these owls commonly produced four eggs in southwestern California and, in the eastern Mojave, even five (P. Bloom, pers. comm. 2006).

Typically, strigid owls produce but one clutch per year, although there are exceptions; for example, in Montana, a Long-eared Owl in a high vole year double brooded, laying one set of eggs unusually early (Marks and Perkins 1999).

Barn Owls are particularly enthusiastic and may produce up to three clutches per year (even more in other countries), with the male sometimes still feeding the young of the first brood while the female is already sitting on a second clutch of eggs at a new nest site (Marti 1992). In southern California, banding records

Fig. 58. A brood of Barn Owls, staggered in size, resulting from the onset of incubation with the first egg laid. (Photo by Steve Simmons.)

Fig. 59. The brood patches of female owls can sometimes be prominent. Northern Saw-whet Owl shown.

have shown that Barn Owls not only produce larger clutches but also produce second or third broods at four- to five-year intervals (P. Bloom, pers. comm. 2006). One Illinois pair using a single nest box, hatched five clutches and fledged young from four of them in a period of 23 months; the young of the failed clutch succumbed during a cold snap (Walk et al. 1999). However, the lifetime reproductive success of 262 Barn Owls studied in northern Utah was surprisingly limited; the mean number of young fledged in a lifetime was a mere 5.58, with 1.03 being the mean number of years of breeding successfully (Marti 1997a).

The eggs are typically laid at intervals of one or two or even more days (although the Burrowing Owl and perhaps other small species may lay more than one per day). Female owls develop a brood patch, an area denuded of feathers on the belly that is heavily vascularized, with the skin thick and glandular and underlain with fat. Incubation begins nearly always with the first egg laid, resulting in asynchronous hatching of the young and the consequent staggering in size, development, and fledging.

Young and Their Care

Newly hatched owls may be nearly naked (as in the Great Horned Owl) or covered in down (as are the young of the Short-eared Owl). They are extremely feeble, and their eyes are closed. For the first five days after hatching, Great Horned Owl chicks cannot maintain a body temperature more than 3°C (5°F) over that of their environment, which may be wintry (Turner and McClanahan 1981). They are said to be poikilotherm, their temperature fluctuating with that of their surroundings and in great need of a provider of warmth.

Fig. 60. Hatched with eyes closed and nearly naked, Great Horned Owl chicks rapidly grow a first coat of white down.

They are brooded by the female almost continuously for roughly one to three weeks, depending on the species. Although usually only females incubate and brood the young, a male Northern Hawk Owl *(Surnia ulula)* was found sitting on a nest containing three newly hatched young and six eggs (Grinnell 1900). Part of the time, the female may be brooding young while still incubating the more recently laid eggs. Typically, brooding females leave the eggs only for very brief periods at dusk and before dawn to cast pellets and defecate, although in some species such as the Northern Pygmy-Owl, she may exit the nest chamber to accept food from the male, leaving eggs or clutches briefly unattended. In some species, and perhaps in all, she eats the fecal material of the young chicks, which may be enclosed in gelatinous sacs like the droppings of nestling passerines (Murphy 1992). She feeds them the food brought by the male, tearing it up and distributing it until the chicks are large enough to swallow prey whole, when she lets them feed themselves, and she now begins to supply food as well.

Prodigious amounts of prey are usually supplied. A male Elf Owl, for example, brought food to the female every two to five minutes at night (Ligon 1968). Barn Owls may deliver as many as six prey items in a little over half an hour, one at a time (Bunn et al. 1982). In California's Central Valley, a pair of Barn Owls during an eight-week period may feed their six young an average of 155 pocket gophers, 113 California Meadow Voles, 20 deer mice, and an additional pound of miscellaneous rodents and birds; all told, the parents and owlets eat about 68 pounds of mostly rodents (S. Simmons, pers. comm. 2005).

Young owls grow rapidly. They may actually come to weigh more than their parents, but lose excess weight before fledging. Parents bringing food alert the young in the nest, and after they leave it with calls, the owlets responding with hisses and other sounds so that they may be located.

Nest camcorders have revealed that owl youngsters soon practice catching prey by pouncing on spots of sunlight, knots in wood, and on branches. The young of hole-nesting species soon begin to scramble up to the entrance of their nest cavity to await food delivery. They, as well as platform nesters, typically leave the nest (and sometimes are enticed with food by their parents to do so) well before they can fly (or at least fly well) and become "branchers." A pair of Northern Pygmy-Owls flew back and forth in front of their nest hole, carrying voles and giving churring calls; within 15 minutes, they enticed their brood to leave the hole, the young exiting at three- to five-minute intervals on their maiden flights (which more resembled dives or crashes into the greenery below). The young perch on limbs near the nest site or amid nearby rocks or even in dense vegetation on the ground. This "prefledging" behavior, typical of owls, appears to be an

Fig. 61. Upon leaving the nest cavity, owl fledglings such as this Northern Pygmy-Owl sometimes wind up hanging upside down from a branch.

adaptation that reduces the risk of having the entire brood destroyed by a predator. Even fledglings capable of gliding are poor at landing and may hang upside down from branches before righting themselves or dropping down to a lower limb; only Northern Saw-whets fly tolerably well when they leave the nest.

Young owls go through distinct phases of development. They learn to socialize with siblings and parents by allopreening or playing with available toys such as rabbit legs and having games of tug-of-war. Owlets raised in captivity reveal much about such kittenlike play practice: a hand-raised young Western Screech-Owl and a Burrowing Owl rolled on their backs while playing with wadded-up newspaper, and they pounced on small balls and shoes. The Burrowing Owl flew onto a newspaper being read by its keeper; when the reader turned a page, the owl flopped on its back and kicked at the paper. When tired, these birds often lay down to sleep and sometimes rested on their sides, lying on one folded wing.

A Northern Saw-whet, picked up in mid-May by well-mean-

Fig. 62. A young Burrowing Owl battles a paper bag.

ing people as a fledgling, became very tame after an initial week's shyness. Turned loose in the evenings in the living room, it would fly about and frequently seek out the human occupants to have its head scratched. When the keepers toyed with its feet, the little owl in return nibbled on the fingers playfully; it also appeared to enjoy landing on a person's shoulder and insinuating itself between the neck and the back of the easy chair to truffle in the hair. The bird quickly learned to return to its cage for food, where it dozed away the day.

After about a month, the young owl gradually began to molt from its juvenal plumage to essentially adult feathering, all except its flight feathers but including those of the facial disk, which changed dramatically. It began to chase moths and dobsonflies *(Neohermes)* released in the room and grew increasingly more skillful at capturing them, even from the undersides of beams, and pursuing them in the air and on the floor. It learned to hover adroitly and maneuvered with great skill and at speed around furniture and other obstacles. It flew in a steep spiral from the floor up to the curtain rods with remarkable ease. It also liked to zoom past people's faces and approach them from behind.

By mid-August, its molt essentially complete (except for the retained flight feathers), the owl looked like an adult Northern Saw-whet. It caught not only insects but also mice released on the floor, dispatching them with bites and great vehemence while mantling fiercely. Although it still uttered its chirping begging calls, its behavior toward its keepers changed dramatically; it became standoffish and no longer enjoyed playing and allopreening. The owl also came up with a dazzling array of novel calls during this maturing period, most of them high-pitched chips and chirps. It, however, never sang the primary song of a Northern Saw-whet nor anything resembling the sound of a saw being sharpened (that being the supposed source of its common name).

Adult owls typically continue to feed their fledged young (which often roost together) for several weeks. The change in appearance, particularly of the facial disk, brought on by the molt of the juvenal plumage into what is essentially adult feathering (the Basic I of field guides) may signal the parents to stop feeding. By late summer and early fall, the young of most North American owls have become independent and disperse, moving away from their parents' territory in various directions.

Dispersal and Migration

The timing of dispersal of juvenile Western Screech-Owls is under the control of corticosterone, a hormone that regulates muscle activity (Belthoff and Dufty 1997).

Young owls leaving their parental lands may turn up in surprising places. Two of six nestling Barn Owls banded in Wisconsin were eventually found dead in Florida, more than 1,900 km (1,200 mi) away (Mueller and Berger 1959). Not quite as adventuresome, 144 Barn Owls out of 2,085 banded as nestlings in Utah, dispersed on average a bit over 102 km (63 mi) from their natal sites, still a substantial distance, avoiding inhospitable areas such as deserts and the Great Salt Lake. Range expansion and the repopulation of former ranges may be helped by such dispersal, which also greatly reduces inbreeding (Marti 1999).

Adult owls may also disperse to breed, from one site to another. Barn Owls, chiefly females, moved to other sites the same year after failing at a first nesting attempt, or in order to produce a second clutch after a successful first nesting (Marti 1999). A Wisconsin adult was captured on a ship 362 km (225 mi) due east of Savanna, Georgia, roughly 1,500 km (940 mi) from its banding site (Mueller and Berger 1959). In the Carrizo Plain National Monument in southern California, eight Burrowing Owls from seven failed nests dispersed; none did from four successful nests (Rosier et al. 2006).

Barn Owls in most of the West are largely or perhaps even entirely sedentary, the Californians particularly, living as they do in a generally benign climate. In Utah, for example, winters can be brutal on this species. Some owl species, especially those feeding on insects and/or those breeding in the high mountains, need to move out when food gets scarce or unavailable in winter. Certain populations of a species may migrate, avoiding the more severe northern winter; this is true for the Burrowing Owl, for example. Other species, such as the Long-eared Owl, travel within California and elsewhere. Movements of this species in the West are not yet well understood, though often they are apparently food dependent (Duncan 2005).

The Flammulated Owl, almost exclusively an insect eater, must head south from its montane habitat to find food. Northern Saw-whets, which in many areas are sedentary, also go south or downslope in colder places, so that they turn up in backyards in the Central Valley in winter and even in desert oases. In the East, where this species migrates in numbers, annual fluctuations in migrants directly reflect breeding success (Cannings 1993). In southwestern Idaho, Northern Saw-whet populations fluctuated with the rodent populations, suggesting that this owl may be nomadic. The Western Screech-Owl population in the same area was unaffected by the decrease in rodents (Marks 1997). If, however, the Northern Saw-whet is in fact nomadic, it does not, unlike other nomads, focus on cyclic prey, such as voles (Marks and Doremus 2000). Another Idaho study found that Long-eared Owls, also thought to be nomadic, were more likely to reoccupy nest sites if they had been successful the previous nesting season (Marks 1986).

Nomadism and dispersal do not lead to an eventual return, unlike migration, which is defined as movement, usually seasonal, away from a breeding area and back. A great deal of information about bird migration has been derived from banding studies, but recent technology advances have added stunning capabilities: migrating Tundra Peregrine Falcons *(Falco peregrinus tundrius),* for example, have been outfitted with solar-powered transmitters containing a global positioning chip, which not only makes it possible to track the exact movements of the falcons but also, by linking the coordinates to Google Earth, to zoom in on their very roost trees in Chile where they winter (B. Anderson, pers. comm. 2007).

Not surprisingly, on migration owls fly by night.

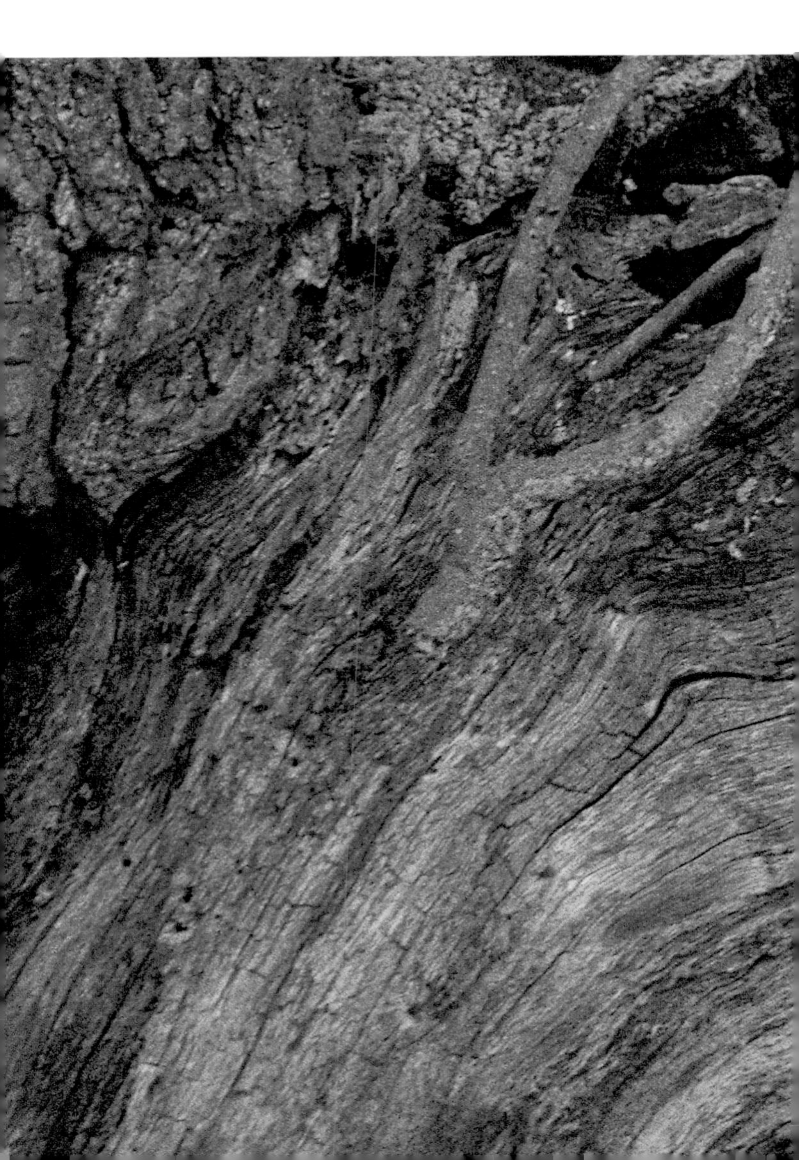

FINDING AND WATCHING OWLS

Finding Owls

A quick look at the distributions of North American owls reveals that the West, and in particular the Pacific states, boasts the greatest diversity of owl species in the country, leading perhaps to the assumption that this region also holds the greatest number of owls. But the person who has actually gone looking for them by day has to wonder just how abundant these birds really are, given how difficult it is to glimpse even one in habitats known to hold several species.

Unlike many hawks, which sit out in the open, most owls are not conspicuous, chiefly, of course, because they are almost all nocturnal in their activities. Only one, the Short-eared Owl *(Asio flammeus)*, might be seen actually soaring like a hawk, though very rarely. The night hunters for the most part hunt from perches, and Barn Owls *(Tyto alba)* are the most likely to be seen on the wing, not infrequently making brief cameo appearances in the headlights of your car.

Finding owls by day is an arcane art that requires not only a sharp eye to discern a lump of mottled feathers nestled against matching bark but also the detective's instinct for picking out the unusual as the searcher takes note of circumstantial evidence. By night, at the right time of year, the task is a bit easier, because owls can be noisy lovers or proclaimers of property rights to other owls.

The temptation is to say that owls are where you find them, because chance so often seems to be the only deciding factor in locating one. However, you can decidedly improve your odds by becoming an owl stalker by night and an owl detective by day.

When to Look for Owls

Although it is possible to find owls year-round, it is definitely easier at certain times of the year. The song of several species is more commonly heard in fall, winter, and spring, leading up to and during the breeding season. Some migratory species augment local populations so that your chances of finding an owl are substantially improved. In late August and early September, before the rains come, visits to woodlands are often rewarding. Because the year's herbaceous growth and grasses have largely been beaten

Fig. 63. In late summer, when the vegetation is down, owl pellets can sometimes be found easily on the leaf litter.

down by cattle and wildlife, spotting whitewash and pellets on the ground is much easier, the dark boluses nicely contrasting with the paler leaf litter. Also, this is the time when molted feathers are found in abundance, and when the young of the year often have not yet dispersed, which increase the chances of actually seeing an owl.

It might seem obvious that the time to look for owls is at night, but some species can be readily located during the day if you know where to look. After dark, many owls can be found by their calls and songs in the first hour after sunset, when they are often particularly vocal and active, and again during the hour before dawn. However, in late summer and early fall, many owls are noisy at three and four o'clock in the morning. Sometimes in the evening there is an orderly sequence of the species announcing their presence, starting with the quite diurnal Northern Pygmy-Owl *(Glaucidium gnoma)* and, after sundown, the Western Screech-Owl *(Megascops kennicottii)*, and Great Horned Owl *(Bubo virginianus)*, then the Northern Saw-whet Owl *(Aegolius acadicus)*, and ending with the blackness-loving Long-eared Owl *(Asio otus)*.

Where to Look for Owls

The Burrowing Owl *(Athene cunicularia)* makes a habit of serenely perching on open ground in broad daylight, and the Short-eared Owl may fly about during the day. The far rarer Great Gray Owl *(Strix nebulosa)* sits in the open and hunts diurnally in winter and when nesting, as does the Northern Hawk Owl

Fig. 64. Canary Island date palms are widely planted in California and are a favorite of certain owls, especially Barn Owls. This Barn Owl, roosting in a Canary Island date palm, has its facial disk in the "at rest" position.

(Surnia ulula), and the scarce Snowy Owl *(Bubo scandiacus)* perches on the ground and on fence posts or other low structures all day long when it visits California and other western states. But generally, owls make themselves as invisible as they can, hiding in dark places and often sitting behind branches that break up their outline. Look for chunky silhouettes in dense trees, although even the large species are masters at hiding.

The hiding places of a few species are occasionally predictable: an owler seeking a Barn Owl does well to investigate isolated or small stands of palms, especially of Canary Island date palms *(Phoenix canariensis)*, and look amid the lowest frond bases. Given the opportunity, never fail to check abandoned buildings (especially in settings with trees) and, of course, barns in the countryside, roosts that are also used by Great Horned Owls and Western Screech-Owls. In prey-rich grasslands, these birds also roost in isolated groves of dense live oaks *(Quercus)*, and in riparian woodland. Migrating Long-eared Owls hide, at times in large groups, in desert tamarisk *(Tamarix)* groves, willow *(Salix)* thickets surrounded by grassland, and small pine *(Pinus)* groves on golf courses. In late winter and early spring, careful examination of old Red-tailed Hawk *(Buteo jamaicensis)* nests in isolated

Fig. 65. The ground under palms is often littered with dates and pellets in various stages of decomposition, and splattered with whitewash.

rural eucalyptus *(Eucalyptus)* groves and elsewhere may turn up the domed, horned outline of a brooding Great Horned Owl, which, at other times may also often be spotted at sundown on the tops of trees against the evening sky. Being large and not very vulnerable, Great Horns may roost on or near the ground, especially

Fig. 66. Western Screech-Owls, along with Barn Owls, sometimes roost in abandoned buildings in farm or ranch country.

in ravines, during hot weather and to escape insect pests (Rohner et al. 2000). Small owls are fond of unused sheds and cabins (with access holes), vine tangles, and the dense foliage that forms at the edges of a live oak's canopy, and Northern Saw-whets hide in clumps of mistletoe *(Phoradendron).*

Because the smaller species of owls are more difficult to locate, you need to first determine their habitat preferences. This information can be found in the species accounts, and the general vegetation distribution map (map 1) will aid in locating the appropriate habitats. Bear in mind that a general vegetation map may show grassland, but riparian zones within that area that are not shown may provide habitat for screech-owls, for example. Next, having determined which owls occur in a given area, you can visit there after dark and listen for calls; in dim light, sometimes it pays to watch for movement in the sky between trees. Excursions during the day may also turn up various circumstantial and direct evidence.

It often pays to inquire with rangers at county, state, and national parks about local owls; not uncommonly, these people are very knowledgeable and helpful and can direct you to a roost site or even a nest. The National Audubon Society's Christmas Bird Count records, birding hotlines of local Audubon chapters, county breeding bird atlases, and Internet birding message boards and listserves are also good places to find productive owling areas. Some owls, such as the Northern Saw-whet, on migration may not stay in an area for very long, but others, such as the nonmigratory Western Screech-Owl, can be found in the same place time and again.

Direct Evidence: Songs, Calls, and Deceptions

Owls resemble songbirds in that they proclaim their territories by vocalizing. The songs or calls of an owl announce the presence of an occupied territory to others of its own kind. These advertising songs of the various species, designed also to attract appropriate partners, are the most commonly heard vocalizations, and many people have come to realize that the bouncing ping-pong-ball song of the Western Screech-Owl or the deep velvety hoots of the Great Horned Owl come from owls. Hearing such songs during late fall, winter, or spring is almost certain evidence that the performer is breeding in the area or is getting ready to do so.

Although both sexes may sing, it is usually the males that begin calling, and females are admitted into territories by answering appropriately. Some owls are sedentary and have permanent pair-bonds and so live together year-round, but they, too, become more vocal as the breeding season approaches. Several owl species perform duets, the male and female singing simultaneously. An owl song can be a varied phrase, such as the "Who cooks for you?" of the Barred Owl *(Strix varia)*, or it may consist of no more than a seemingly endless monotonous series of "toots," the song of the Northern Saw-whet Owl.

For serious night owling, the purchase of recordings of the songs or calls of North America's owls is decidedly a worthwhile investment. The immediate payoff is the ability to associate these vocalizations with the singers and, even better, to learn to imitate them. Most owls are highly territorial and readily respond, at least during the breeding season, to the whistled or voice-produced imitation of their advertising songs even when the copy is far from perfect. It is a highly satisfying experience to get a dialogue under way between yourself and an owl. Most species respond with considerable ardor, as revealed when you, having followed the respondent's voice, discover in the flashlight's beam an agitated creature with glaring eyes and throbbing throat as it throws out one challenge after another to the perceived interloper.

Occasionally, an owl, half asleep in its hiding place, may softly answer an owl seeker's probing whistle during the day, sometimes repeatedly, providing a beacon for the searcher, its presence additionally given away by pellets and specks of whitewash under its perch. However, an owler who returns to the same area after dark is likely to be surprised by the decidedly more enthusiastic response to whistle or hoot and by the vocal flourishes and wails added by the now wide-awake bird: for most species, nightfall is morning.

Considering how hearing-oriented owls are, it is curious that the exact pitch of a song's imitation is not all that critical; more puzzling still is the vocal response of one species to the whistled or hooted copy of another species' call, particularly when they are about the same size and one is not known to prey on the other. Some owls may fly toward the imitator and perch nearby, twisting and craning their necks and bobbing their heads as they try to determine where exactly the interloper is hiding. Worked-up Great Horned Owls, well known for a take-no-prisoners attitude, may install a new part in a person's hair. Such a large owl looking for a trespasser in its territory is no laughing matter, especially in the vicinity of its nest, where the intruder, in the owl's mind, may represent an immediate threat to its young. Northern Pygmy-Owls often approach the mimic but may then refuse to show themselves; however, an enthusiastic Northern Saw-whet landed on the binoculars hanging from one owler's neck (P. Gordon, pers. comm. 2005). It should be remembered that in many parts of the West, including California, there is the added thrill of a possible nocturnal encounter with a Mountain Lion *(Puma concolor)*, so it is wise to go with a companion for moral support and to form a defensive unit.

Imitating the song of a small owl will sometimes cause a larger species to fly in at speed in hopes of making a meal of the caller. Perhaps because of this, the songs of the small species have a remarkable ventriloquial quality, often making it very difficult to locate the source or even accurately assess the distance to it. Following the tooting of a Northern Saw-whet Owl well after nightfall, a pair of hopeful owl watchers initially could not decide whether there was one owl or two calling back and forth, the second about 50 yards away from the other. After much stumbling about in the black woods, the searchers felt they were close to at least one of them and, turning on the flashlight at last, found the bird perched about 2.5 m (8 ft) above their heads. To their astonishment, the owl could greatly lower the volume of its voice and, by turning its head to the side, could project its calls so that they seemed to come from a considerable distance away; there was, after all, only one owl.

Hooting like a big owl may, on the other hand, bring in small ones intent on mobbing the larger predator. Western Screech-Owls, for example, may approach, sometimes two or three together, with a cascade of peevish sounds. Elf Owls *(Micrathene whitneyi)* respond enthusiastically to their recorded advertising song during breeding season but respond well to the call of a Great Horned Owl year-round (Boal and Bibles 2001). Generally, however, when testing a new and promising area, it is advisable to offer imitations of the smaller species first, lest they be driven deeper into hiding by your menacing Great Horned Owl or Barred Owl hoots.

Keep in mind that an owl might not respond immediately or at all. The art of owling by ear is further complicated by the perplexing variety of calls produced by a given species, sometimes making identification difficult. For example, just prior to copulation, a female Northern Pygmy-Owl under observation by the author let out a cackling staccato call of such volume and pitch as to seem wholly incompatible with such a tiny bird, a shriek I have not heard since; perhaps it was the "chatter call" mentioned in the literature (Holt and Petersen 2000), in which case it would illustrate the difficulty of accurately transcribing some of these vocalizations.

Owls also produce a great variety of other sounds. A Great Horned Owl can emit calls that sound like those of the much smaller screech-owls, and the latter can conversely bellow with amazing volume. Generally, imitating such vocalizations does not lead to continuing conversations that would help in tracking down the caller, but an attempt should be made to follow the sound. Sometimes the source is completely unexpected.

Sometimes it is possible to identify the species by the contact and begging notes of juveniles, as in the case of the distinctive raspy notes of the young Great Horned. A young Northern Saw-whet, however, produced a bewildering variety of calls, the function of some unfathomable. Some, chirps resembling those of a songbird, were clearly begging calls. Young Northern Pygmy-Owls, however, beg with notes so similar to protracted sequestration calls of a Bushtit *(Psaltriparus minimus)* that in daytime it pays to investigate the source of such sounds.

It is of course also possible to take along and play recordings with a portable player of some sort when one goes afield owling; loop tapes (like those used in answering machines) make locating a particular call easier. iPods, which give quick access to as many calls as most birders could possibly desire, may soon make tape players obsolete, although at present they cannot be used for recording calls in the field because of intentional blocks on the software to prevent recording of some frequencies, for copyright infringement reasons. Commercial downloads of bird calls are readily available.

However, such artificial taunting should only be a last resort and is in fact banned in some parks. Playing of recordings by day has been known to attract not only owls but also hawks that may kill the owl (an effect that has been especially troublesome with

Spotted Owls [*Strix occidentalis*]), and such playing by night may draw in two owl species, with the larger attacking the smaller (B. Woodbridge, pers. comm. 2006). Excessive use of tapes by eager birders has been known to drive owls from their territories or even cause them to abandon nests; at the very least, it causes them to waste much energy in responding needlessly to a perceived threat or competitor (see also "Hazards to All Owls" in the section "Owls and Humans"). Amusingly, there are numerous instances where one owler's imitations were answered by another's, in at least one case almost nightly for several months (J. Loft, pers. comm. 2004).

Squeaking (reproducing the sound of a small rodent in distress by pressing your pursed lips against the back of the hand and sucking) can also be effective in attracting owls. Enthusiastic respondents have been known to strike the squeaker in the head, however.

Barn Owls and Great Horned Owls are often frightened off by the watcher's noisy progress through the woods, but others, such as the Northern Pygmy-Owl, are not. Move quietly, and if you squeak, make an effort to conceal yourself, although that often seems unnecessary.

Circumstantial Evidence

Although owls are intensely private creatures, they cannot avoid providing evidence of their whereabouts; sometimes their presence is given away by droppings, remains of meals, molted feathers, eggshells and by informers, such as songbirds that may noisily mob an owl.

Whitewash and Pellets

Named after old-fashioned house paint, whitewash splashed on cliffs, forest floors, trees, or pavement or caked on branches indicates the presence of meat-eating birds, including owls, hawks, and some nonraptors. Like all carnivores, owls produce great quantities of nitrogenous wastes resulting from protein digestion that are eliminated by the urinary system and that constitute the white part (the uric acid) of any bird's two-tone dropping. The dark part is digestive waste from the intestine. In small owls, this component is often shaped like a small worm, whereas in large species it is sometimes broken up and mixed with the white.

Fig. 67. The dark digestive wastes in the droppings of small owls are distinctively worm shaped, as in these Burrowing Owl deposits.

Fig. 68. The tunnel (or "form") of a Short-eared Owl. Note the pellets and rather solid droppings.

Short-eared Owls' digestive wastes look like soft-edged, dark buttons sitting atop the creamy white uric acid, the whole dropping resembling a fried egg. The excreta, together with a few pellets, are often found in the short tunnels made by this owl in tall grass. Generally, compared to hawks' whitewash, that of owls is more crumbly, denser, and often flaky and creamier in color.

Because owls may use the same roost for long periods, even decades, whitewash accumulates below. In fact, some perches are so desirable that they are sought out by one generation of owls after another, and in some cave sites, the buildup of whitewash, protected from the elements, is such that it suggests continual use over centuries if not millennia, typically by Barn Owls. Surely it takes very little detective work to locate Barn Owl roosts decorated in this fashion, and even those of Great Horned Owls. Such painted rocks are most easily spotted in the desert and the Great Basin areas, where rain does not wash the whitewash away.

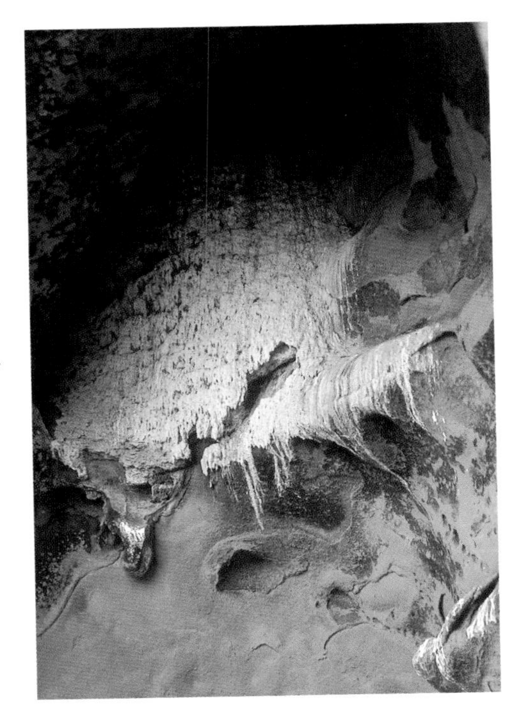

Fig. 69. In caves, protected from rain, whitewash buildup can look like (and probably is) the product of millennia of Barn Owl droppings.

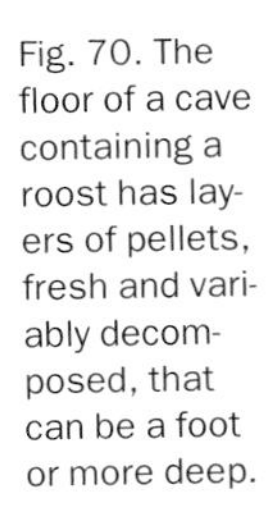

Fig. 70. The floor of a cave containing a roost has layers of pellets, fresh and variably decomposed, that can be a foot or more deep.

Fig. 71. On the cliffs of deserts and the steppe country of the Great Basin, the whitewash of owls, chiefly Barn Owls, can be seen from afar.

A few pellets of undigested fur and bones usually accompany the whitewash beneath a roost; traditional sites have great heaps of such ejecta, with most having decomposed into a kind of humus speckled with bones. Only relatively few recently produced, intact, charcoal black pellets may lie on top.

A number of other birds produce pellets, too, an important fact to remember. Loggerhead Shrikes *(Lanius ludovicianus)*, for example, eject very small ones, smaller than those of any owl, but a number of diurnal raptors make pellets similar to those of owls except that, with the exception of those of the White-tailed Kite *(Elanus leucurus)*, they contain very little, if any, bone.

A small pellet found in the woods is not necessarily the product of a small owl species or a baby owl (which produces pellets typical for its species early on) but may simply represent the single-mouse meal of a midsized or large species, and a small owl may eject a tiny pellet from eating only the head of a mouse.

Any amount of whitewash noticed on a walk through the woods should be cause enough for you to look up into the tree overhead. Perhaps it only indicates an owl's regular hunting perch overhead, and most often you will see only leaves and branches; but sometimes a round and earnest face gazes back.

Feathers

A pile of owl feathers on the ground is postmortem proof that the species lived here until it was eaten by another predator, such as a Bobcat *(Felis rufus)*, a hawk, or another owl. At the least, such remains identify a habitat for that species.

Conversely, a pile of feathers, found in the woods, of an American Kestrel *(Falco sparverius)* or other open-country bird suggests that the bird was at its night roost where it likely fell victim to an owl. Pulled-out wing and tail feathers with bite marks appropriate in size for a Great Horned Owl's beak can confirm this suspicion.

A hiker in the woods not uncommonly comes across a molted owl feather that catches the eye with its intricate pattern of earth colors. Because in at least some owls the molt of the all-important flight feathers can take more than a year, you do not find nearly complete sets of shed primaries in the woods as you sometimes can near the nests of certain hawks. Body feathers, especially the large, banded ones of Great Horned Owls and the smaller spotted ones of Barn Owls do turn up regularly in summer and fall, and, coming from owls, they are exceptionally fluffy and lightweight and catch up on stems and stalks, where they are as conspicuous as little flags: an owl lives here!

Egg Shells and Headless Mice

When owlets hatch, their eggshells are either eaten or removed by the parents. Some drop them right outside the nest hole; others carry them away, usually no great distance. Owl eggs are plain white and generally more spherical than the eggs of other birds, a trait still evident in shell halves; such a shell fragment is a sure indicator of a nearby owl's nest.

Owls also betray their presence by frequently (and apparently accidentally) dropping decapitated mice, rear halves of mice, or entire rodents under their roosts and near their nests. Under one Barn Owl nest, 52 whole rodents were found: 42 California Meadow Voles *(Microtus californicus)*, eight deer mice *(Peromyscus)*, one House Mouse *(Mus musculus)*, and one kangaroo rat *(Dipodomys)* (Dixon and Bond 1937).

Fig. 72. An observant hiker in the woods not infrequently finds mouse halves, such as this Deer Mouse *(Peromyscus maniculatus)* rear and California Meadow Vole front end, dropped by careless or inept owls.

Diurnal Raptors

The presence of some hawks may indicate that certain owls are nearby. Northern Harriers *(Circus cyaneus)* and Short-eared Owls like the same habitats and foods, as do American Kestrels and Western Screech-Owls.

Should you happen to find the nest of a Cooper's Hawk *(Accipiter cooperii)*, it is worth your while to search for old nests of that species, which are often nearby and are sometimes used by Long-eared Owls. Great Horned Owls frequently avail themselves of the old (or surrendered) platform nests of Red-tailed Hawks, and the groves and windbreaks of blue gums *(Eucalyptus)* seen so commonly in central and southern California very often are home to nests of both species.

East of the Rocky Mountains, Barred Owls are considered the nocturnal avian predator equivalent to the diurnal Red-shouldered Hawk *(Buteo lineatus)*, with both occupying the same habitat. Having entered northern California in recent years, this owl species is currently expanding its range in the state, as it has to the north. It remains to be seen whether it will seek out Red-shoulder habitat here as well; currently it seems restricted to coastal and montane coniferous forests, although an individual was recently (2006) observed in oak-bay woodland in the San Francisco Bay Area just north of the Golden Gate Bridge (S. Stender, pers. comm. 2006).

Follow the Mob

The cacophony of birds mobbing something in the woods must always be a powerful attractant to the owl watcher, as inconvenient as the investigation might be—all too often it involves poison-oak *(Toxicodendron diversilobum)*. A pack of buzzing and ticking songbirds of several species in a bush raises the possibility of a beleaguered Northern Pygmy-Owl, a species often hard to find but particularly loathed by small birds, doubtless because of its songbird-eating habits. An added benefit for the lucky human coming upon such a riot is the opportunity to view close up some beautiful, normally elusive songbirds. One such owl in Alameda County, California, in early October attracted a lively mob that included Anna's and Allen's hummingbirds *(Calypte anna, Selasphorus sasin)*, Nuttall's Woodpecker *(Picoides nuttallii)*, Oak Titmice *(Baeolophus inornatus)*, Chestnut-backed Chickadee *(Poecile rufescens)*, Ruby-crowned Kinglets *(Regulus calendula)*, Black-throated Gray Warbler *(Dendroica nigrescens)*, Townsend's Warbler *(D. townsendi)*, Spotted Towhee *(Pipilo maculatus)*, California Towhee *(P. crissalis)*, and Golden-crowned Sparrow *(Zonotrichia atricapilla)*. All of these were so intent on harassing the little raptor that they completely ignored human observers just a few feet away. Nearby Steller's and Western Scrub-Jays *(Cyanocitta stelleri* and *Aphelocoma californica)* did not seem interested until the owl flew, causing a few of them to fly after it.

In April, European Starlings *(Sturnus vulgaris)* drew attention to a perched Northern Pygmy-Owl by giving predator alarm calls while mobbing it, perhaps because their young in nearby tree cavities were vulnerable to this known nest robber (see "Mobbing" in the section "An Owl's Life" for a description of the calls and fig. 45). Similarly, other nesting passerines, diving time and again at a small lump high in a tree, are almost certainly harassing one of these tiny owls, and crows *(Corvus)*, plunging likewise into a leafy tree while voicing their displeasure, could be harassing a Great Horned Owl. Alarmed towhees, American Robins *(Turdus migratorius)*, and jays may orbit a Western Screech-Owl in a dense California bay *(Umbellularia californica)*. On the other hand, such commotions may reveal no more than the neighbor's cat or a garter snake *(Thamnophis)*, or, most annoying of all, nothing whatsoever, with the birds either having given a false alarm or with the danger having moved on. Or maybe you simply failed to locate the little owl.

Mammalian Indicators

Beechey Ground Squirrels *(Spermophilus beecheyi)* very often share the same habitat with Burrowing Owls, which avail themselves of the rodents' burrows for nesting. The colonies of these squirrels, easily recognized from a distance by their holes and the grazed-down surrounding vegetation, can be quickly checked for the presence of the owls.

California Meadow Voles make their presence known by their narrow-gauge tunnels, runways, and holes in usually dense grass, evidence of their presence especially noticeable when the populations peak cyclically. An abundance of voles attracts Short-eared Owls, sometimes to nest in spring in such a prey-rich area, but particularly in winter when flocks of Short-ears roost and hunt in such fields.

Fig. 73. A peak year for meadow voles is an important event for owls and is readily detected from the numerous vole runs in grassland.

Alfalfa fields dotted with the mounds of Botta Pocket Gophers *(Thomomys bottae)* are certain to attract Barn Owls by night; the fan-shaped mounds of loose soil are easily recognized by their dirt-plugged holes.

In coastal forests and oak woodlands of California and the Cascades of Oregon, Dusky-footed Wood Rats *(Neotoma fuscipes)* give away their presence by the bulky nests they build both on the ground and in trees; these rats are a favorite food of the Spotted Owl and others. Conspicuous white and tarry encrustations on rocks in much of the dry West are from the urine of the Bushy-tailed Wood Rat *(N. cinerea)*, another species much favored by owls.

Finding Owl Nests

If owls are difficult to find, locating the nests of most is, in most cases, even harder. The species that nest in tree cavities generally provide little external evidence; with luck, there may be a bit of whitewash on the ground directly under the hole (or nearby), or even a molted feather or a headless mouse. On cold, sunny days, an owl may be found sitting in its nest hole, where only a very keen eye will notice it. Usually, however, there are a great many possible holes in a given area, and to pick out an occupied one normally requires night visits and plenty of patience, although scuffed bark or wood around it, caused by owl traffic in and out, sometimes is a clue. The nests of a number of species can be found by listening for the begging calls of the young after dark (P. Bloom, pers. comm. 2006). During the day, flies entering a tree hole often indicate an occupied owl nest holding decomposing food remains. A female Great Horned Owl may call from the nest.

Most breeding diurnal raptors and their young molt copious amounts of white down that conspicuously clings to their nests, but owls do not provide such obvious advertisement.

Stick nesters also do not necessarily select the most obvious old Red-tailed Hawk mansion available but may choose instead the ragged, half-collapsed remains of a tree squirrel *(Sciurus)* nest (recognized by its mix of twigs and dead leaves) that one might not deign worthy of a second glance; often, by the time the young leave the nest, it is in ruins. Long-eared Owls seem fond of old crows' and Cooper's Hawks' nests, which are often hidden in tree crowns and not readily seen from below.

In northern California and elsewhere, nest boxes for Wood Ducks *(Aix sponsa)* have been placed on trees and posts near water in order to enhance the numbers of this hole-nesting species by increasing availability of nest sites. It is often possible to detect which boxes have been commandeered during nesting season by Barn Owls, because the lower lips of the entry holes may appear greasy from numerous rodent corpses being dragged across the threshold, and not uncommonly there is a powerful smell of rotting mice. In general, Barn Owl nests are likely the easiest to find because the birds are numerous and readily accept nest boxes in agricultural areas and flat spots under the roof inside barns and other buildings, as well as conspicuous Common Ravens' *(Corvus corax)* nests built into cliff niches and potholes.

Fig. 74. A Northern Saw-whet Owl used this box put up for Wood Ducks, raising the possibility that the maiden "flight" of the young might carry them into the water. (They can swim, however.)

Western Screech-Owls and Northern Saw-whets, too, will sometimes move into Wood Duck boxes or boxes put up especially for them or for American Kestrels. Normally, however, the nests of these small owls are hard to find, with that of a pygmy-owl even harder by an order of magnitude because of the diminutive size of that species—a flying one is usually lost from sight as easily as a songbird and may vanish from view instantly by shooting straight into a tree hole.

Only the presence of a Burrowing Owl's nest can be surmised with some ease, because of the adults' habit of frequently deco-

Fig. 75. A Burrowing Owl perched high on a cable, telephone wire, or telephone pole is usually an indication that there is a nearby nest.

rating the ground surrounding the burrow opening with rubbish and of perching next to the hole or on a nearby fence post and, later, by the young appearing around the mouth of the burrow. Sometimes, the presence of a nearby Burrowing Owl nest can be inferred from observing an adult perched on a cable or a wire between telephone poles during the day, as it performs sentry duties. A fencepost decorated with whitewash and pellets at its base and, occasionally, a few molted feathers is surely a favorite perch of an owl or a hawk.

Great Horned Owls living in parks have, at times, become so accustomed to humans that they may nest in plain sight, using niches in small cliffs or dirt banks or platforms especially provided for them in trees. As with other owls, their young prefledge: they leave the nest before they can fly and clamber about on neighboring branches, ledges, and such. During this period, they often become quite conspicuous. It is wise to remember that there is always an adult owl nearby who may take the most severe measures should it consider its young threatened.

Frequently, the presence of owls' nests is made obvious after the fact: once owlets leave the nest, they become much noisier to insure the food-bearing parent can find them in the dark. The peculiar growls, buzzes, shnarks, and chatters of the various species are very distinctive and are often not recognized as coming from owls, again offering opportunities for surprise encounters for the adventurous owler.

Watching Owls

Few outdoor experiences can match in charm the doings of a family of Burrowing Owls once the young are large enough to spend most of their time above ground. Certainly no other owl species so readily allows humans to observe its home life. Not only do these birds show us the courtesy of living their lives in broad daylight, but, with a little patience on the observer's part, they soon come to all but ignore a car parked nearby that serves as a convenient blind. In addition, their nest sites are very often prairie dog *(Cynomys)* or ground squirrel *(Spermophilus)* holes; the latter are frequently found at the edges of roads and on levees and other disturbed ground, making the owls not only fairly easy to locate but also easy to approach by vehicle.

Once the observer has settled in about 20 feet away, perhaps with appropriate music on the car radio, both the adults and the half dozen or so owlets soon stop staring statuelike at the scarcely hidden intruder and behave like normal Burrowing Owls, that is, like very cleverly made animated wind-up toys. The young stretch their wings or lie down or sprint hither and yon or nibble each other or try to pick up objects from the ground or turn their heads upside down in the most appealing way. They may engage in mutual head-preening sessions or with one of the parents, showing facial expressions that, to a human, denote consummate bliss. They may rise straight up into the air on new wings, helicopter style, and land clumsily. Meanwhile, one or both parents doze by the burrow standing on one foot, or one or the other may hop or run after a passing earwig, which is soon deftly snatched and eaten or passed on to a youngster.

Although the adults provide most of the food at twilight, some feeding also occurs throughout the day. Sporadically, one of the parents goes off to hunt and returns with prey such as a cricket that soon becomes the hapless object of a vigorous tug-of-war. Even the slow-motion blink of the drowsy owlets is entertaining to watch, as is their preening, and the observer soon learns that when the family gazes as one up in the sky, it is likely signaling the passage of a Red-tailed Hawk or other raptor overhead.

When Barn and Great Horned owls nest in fairly exposed potholes in cliffs or in human-made structures, the young may be watched from a distance; but the intimacy of a visit with

Fig. 76. A pair of Burrowing Owls intently watches a passing Red-tailed Hawk, while the young seems unconcerned.

Burrowing Owls is lacking, and the young are usually much less active, with the adults normally staying out of sight.

An owler wishing to watch the home life of one of the smaller forest species must go out after dark to witness (after having followed the begging calls and with the judicious use of a flashlight) the delivery of prey by a screech-owl to its excited young. Military night-vision goggles may also help, but they are not only very expensive but heavy as well. A more convenient albeit secondhand experience is provided by organizations that install "critter cams" to record the activities inside nest boxes put up for owls—the gulping down of entire mice, the breaking up of pellets to make bedding, the tender way in which the female feeds tiny bits of meat to her newly hatched young, the play behavior of the growing owlets. To see for yourself, try http://www.owlpages.com/links.php?cat=Owls-Nestcams.

Obviously, it is because most owls are resolutely nocturnal that they are so difficult to actually watch. It is, however, possible to observe a very few species during the day. Normally, an owl discovered by day—one that has not been frightened into flight—might for all intents and purposes be a stuffed one as it waits

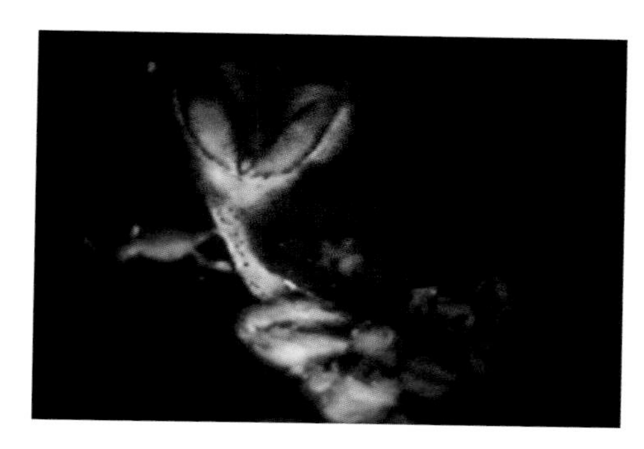

Fig. 77. A critter cam installed in a Barn Owl nest box allows humans to observe the owl's home life on a monitor.

nearly motionless for you to just go away. But along with the Burrowing Owl and the largely diurnal Northern Pygmy-Owl, the Short-eared Owl often hunts while it is light, as do the Northern Hawk Owl, Great Gray Owl, and Snowy Owl, and it is a delight to watch these mysterious creatures.

Owl Photography

Compared to photographing diurnal raptors, taking good pictures of owls is child's play—provided you can find the owl in the first place and have the right equipment. With the exception of the Great Horned Owl, which can at times be a bit skittish, most owls tolerate reasonably close approach as long as eye contact is avoided.

A moderately long telephoto lens, such as 300 mm or even shorter (for film cameras) suffices to take satisfactory pictures; and in poor light, the photographer can set up a stabilizing tripod without fear of spooking the subject.

The advent of digital photography has been an enormous boon to taking great pictures of birds. Although the most advanced film cameras virtually run themselves when they are pointed in the right direction, by selecting appropriate aperture and shutter speeds and even keeping moving objects in focus, good digital cameras do all of that and have an additional advantage: film speed (ISO)—which affects, among other aspects, the graininess of a picture taken with a conventional camera—is no longer a concern, because the ISO number can be changed *after* taking the image. Although some cameras can be fitted to a tripod-mounted spotting scope, excellent pictures have been obtained by simply holding the digital camera's lens against the scope's eyepiece. Even inexpensive camera models have produced stunning results. Photos in which birds appear hopelessly tiny on the digital camera's LCD monitor can be cropped, in effect enlarging the desired portion of the picture, which, if the camera has been preset to shoot at high resolution and has sufficient megapixels, may reveal a clean, crisp image of the bird even when printed large. In addition, the photographer can keep clicking away at a moving subject and can later erase from the camera's memory card the pictures that have no appeal, thereby preventing the accumulation of unwanted slides and prints in cabinets at home.

Like the use of tape players, excessive photographic activity, especially with the use of flashes, is not conducive to a successful nesting attempt by an owl. For the good of the bird, even an avid photographer must exercise self-restraint around nest sites.

Infrared video camera systems facilitate the nocturnal study of owls. These cameras, installed near owls' nests, together with supplemental infrared light sources to extend the instruments' vision, can record details and behavior of such quantity and quality that would be impossible to match by direct observation (Delaney et al. 1998), and they cause less disturbance for the birds.

Owl Identification

Most North American owls are fairly easy to tell apart if you focus on a few prominent features and take the habitat into account. Two sometimes very visible species, the Burrowing and the Short-eared, are found nearly always in open areas, where they may be observed during the day. The Barn Owl and the Long-eared Owl also like open places for foraging (the Long-ear a bit less so) but are not averse to the presence of trees, and they are almost exclusively nocturnal. The majority of owl species likely to be encountered in North America are, however, rarely found away from trees.

The chief physical traits that identify owls are the size, from diminutive to very large, the eye color (and, to a lesser extent, eye size), the shape of the body, head and facial disk, and the presence or absence of ear tufts. These structures may be conspicuous or difficult to see; even species with large tufts can temporarily flatten them against the head, and the tiny Northern Pygmy-Owl, which has no obvious tufts, on very rare occasions can raise up a pair of minute tuftlets. Northern Saw-whets and Boreal Owls *(Aegolius funereus)*, when attempting to conceal themselves, lift the upper outer "corners" of their facial disks so that they form blunt projections somewhat similar to tufts.

Fortunately, most owls tolerate fairly close approach, enabling the observer to accurately assess size by comparing it to objects near the bird, such as a leaf. Size however, can be deceptive, for instance if the owl is perched high in a bare tree, where its outline against the sky in dim light makes it appear larger than it really is. Light-colored owls, such as Barn Owls, to many people famously appear very large in the headlights as they flare out of the way of the car.

Eye color, in most cases, is easy to ascertain; even when an owl is pretending to be a branch stub, with its eyes reduced to narrow crescents, it is still possible to make out their color with a good pair of binoculars. There are two possibilities: the eyes are very dark brown, so as to appear black, or they are some shade of yellow, from pale gray yellow to brilliant orange red. The juveniles of some species have oddly cloudy eyes, the pupils appearing bluish.

An owl's shape is occasionally a good clue, too. A bulbous, barrel-shaped body the size of a professional football fitted with a grapefruit-shaped head belongs certainly to either a Spotted Owl or a Barred Owl, even when seen only as a silhouette under the forest canopy. Both Long-eared and Short-eared owls have rather pointy heads, with the ear tufts arising not too far from the midline. The Barn Owl's facial disk is uniquely heart shaped, and the head of the Northern Saw-whet looks often like a rectangle with rounded corners, or even trapezoidal, so that the little owl looks as if it were wearing a medieval lady's headdress. The only other owl with that head shape is the Boreal Owl. It must always be remembered however, that owls can change the shape of the head and especially that of the disk by raising and lowering the feathers or with the aid of muscles underlying the skin. Although screech-owls and the Northern Saw-whet are very close in size and all have yellow eyes, screech-owls can muster up expressions of fierceness, whereas Northern Saw-whets always look gentle or startled.

An owl's voice is often considered its calling card par excel-

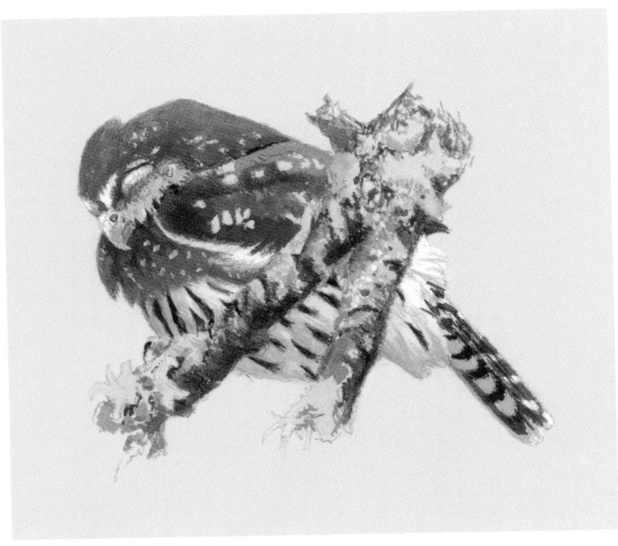

Fig. 78. A young Northern Pygmy-Owl sleeping on a branch would be difficult to key out, but the long tail and the size give it away.

lence. Certainly the advertising songs of most species are very distinctive and provide definitive identification even in the absence of visual clues; however, each species has, in addition, an extensive repertoire—often over a dozen different calls and songs—and many times, unless the caller is seen or offers its advertising song as well, identification can be an uncertain business. For example, the voice is critical in separating the screech-owl species where their ranges overlap. The songs of both the Mourning Dove *(Zenaida macroura)* and the Band-tailed Pigeon *(Patagioenas fasciata)* are sometimes mistaken for owl songs; both species, however, vocalize exclusively during the day, which is very unusual (but not unheard of) for an owl. Day-singing, when it does occur, is often a softer version of nocturnal efforts.

Behavioral observations can greatly aid identification. If at rest, where is the owl perched? If in a tree, is it up against the trunk or away from it? If in an open area, is it sitting on the ground, on a fence post, or even on a telephone wire? Does it appear awake or asleep? If flying, is its flight undulating, like that of a woodpecker? Does it fly high above fields in the day, or even soar? Or hover? Does it call while flying? The answers to these questions can contribute to the identification of an owl.

OWLS AND HUMANS

Attitudes Past and Present

Owls have infused the art, myths, and literature of humans since prehistoric times. Paleolithic cave paintings in France dating back to 15,000 to 20,000 B.C. include a depiction of two Snowy Owls *(Bubo scandiacus)* with their chick, as well as some sort of horned owl. In ancient Egypt, the owl pictograph stood for the "m" sound, and the ancient Maya utilized owl glyphs also. The Greek goddess of wisdom, Athene, is often shown in the company of an owl. The perception that owls are wise persists to this day, an idea bolstered perhaps by their deliberate behavior and unwavering gaze. Alas, compared to other birds, owls are not particularly intelligent and are slow to learn even simple tasks compared, for example, to hawks and parrots.

Whatever the reason, an owl appears on L. Ghiberti's famous bronze doors of the Florence baptistry dating back to the early Renaissance. In Goya's painting *Don Manuel Osorio de Zuñiga,* the raptor is shown as a caged attractant, used to lure and trap small birds. This method of "hunting" is mentioned by Aristotle and was still practiced legally as recently as 1992 in Italy, where other forms of songbird hunting continue to this day, the little corpses being then consumed with polenta and a nice chianti.

The Greeks ambiguously thought of owls as evil omens or portents of victory. An army composed chiefly of Greeks fighting the Carthaginians in the third century B.C. was much encouraged by noting these birds perching on their banners, unaware that their commander had ordered some captive owls released at just the proper moment (Armstrong 1975). The Tartars wore eagle-owl feathers in their caps to ensure success in war, because, as the legend goes, an owl sat above the spot where their leader Genghis Khan was hiding, and the enemy passed by without searching for him, believing that an owl would not sit so near a man (Loyd 1927).

A half dozen owl species are identifiably mentioned in the *Bible*, all of them considered unclean because they are meat-eaters. Shakespeare made free use of their symbolic value; for example, his witches in *Macbeth* add an owlet's wing to their cauldron. Witches were sometimes depicted as owls, because they were thought to transform themselves into owls at night. Chaucer thought owls ominous, an opinion unfortunately shared by a

Fig. 79. An owl, presumably the Little Owl (*Athene noctua,* an Old World species), carved from wood, has movable wings. This decoy is used by Italian songbird hunters to attract their quarry. (Photo by Daniele Colombo.)

majority of people worldwide. In parts of the globe that as yet have no electricity, the darkness is still filled with terror where ghosts and wild animals shriek and lurk and ambush sightless humans, and the screaming phantasm of a passing owl provides the perfect embodiment of this primal fear.

On the other hand, a saying in Sussex, England, went, "When owls whoop much at night, expect a fair morrow," and agricultural societies probably soon realized that at least the Barn Owl *(Tyto alba)*, which so readily takes to living with humans, was in fact a valuable ally in the protection of stored grain against rodents. This did not, however, eliminate ancient superstitions that permeate even modern urban societies. On the next visit to your favorite ethnic restaurant, ask about the attitude toward owls in a foreign-born proprietor's home country. Chances are the response will be negative: hearing an owl's call means your business deal will fall through, or you will be taken ill, or worse. Rarely do owls foretell happy events.

In addition to augury, owls also provide their bodies or parts thereof for human uses. In China, for example, owl soup is not only a sought-after dish but is also thought to have medicinal properties for curing consumption and rheumatism (Everett 1977), and in Morocco, a wife or daughter will be forced to tell the truth about her daytime activities if the right eye of a Eurasian Eagle-Owl *(Bubo bubo)* is placed in her hand as she sleeps (Marks, Cannings, and Mikkola 1999). In Germany, owls were nailed to barn doors to avert lightning.

Among North American native peoples, too, owls have been subjects of superstition, and in some tribes they were considered both good and evil. Some stories tell of owls carrying away children to eat them, (Marks, Cannings, and Mikkola 1999), and in many tribes, owls were linked to imminent death. They often were regarded as conveyors that took souls to an afterlife. Yet in the Dakotas, a little owl (probably the Burrowing Owl *[Athene cunicularia]*) that flew above a Hidatsa warrior riding to war was perceived as a protective spirit (the warriors also wore owl feathers for protection), and a big owl (likely the Great Horned *[Bubo virginianus]* or Snowy controlled the movement of Buffalo *(Bison bison)*, which the owls kept seasonally confined in an enormous butte (Johnsgard 2002).

Native Californians and Owls

In California, it was common practice among the various tribes to tell stories and myths for evening entertainment, and some of these featured anthropomorphic owls that acted like humans, conversing and interacting with other articulate animals and with people.

A tale told by the Yokuts (San Joaquin Valley) involved a man who, having attracted and shot an owl for dinner, at the urging of his wife called in and shot so many more that he ran out of arrows. But the owls kept coming and, in the end, killed the couple (Margolin 1993). This lesson in wildlife conservation is matched by a morality tale of the Wiyot tribe of Humboldt County, of an owl man who grew fat on meat he hid from his starving wife and children. Having found him out, the enraged wife banished him to a distant dark cave, "and that is where all owls live, away from everyone, by themselves, talking to themselves" (Gifford and Block 1930).

If a Yokut hunter heard a sharp whistling sound coming from

a Pronghorn *(Antilocapra americana)* as it ran past, he would not shoot it, because it was believed that a Burrowing Owl, when it died, would select a particular antelope to guard and that the owl's ghost was making the sound. In that culture, the Burrowing Owl was the representative of the professional singers, who performed at the lonewis, the annual ceremony for the dead (Latta 1949).

The Lake Miwok (Lake County) considered the "Little Owl" a great doctor (Gifford and Block 1930). However, one Costanoan (Bay Area, also known as Ohlone) Indian, when prophesying death, was recorded as saying, "Watch the owls. When they come in the night, they will have no voice," and he also warned, "You watch when the owl goes like this [he imitated a call], then you know it is bringing sickness" (Collier and Thalman 1996, 494). Ohlone shamans might wear a stuffed owl around the neck when trying to cure a person of sickness; they had the power to send an owl to frighten an enemy, but they had to be watched, because they could turn evil and learn to communicate with owls (Margolin 1978).

Yuroks (Humboldt County), suspecting that an owl's call might really be made by a devil imitating an owl, would stop and build a fire, to show the devil that they would not be taken unawares (Thompson 1991). The Sierra Miwok believed that when good people died, they might turn into Great Horned Owls, but bad people would become a Barn Owl, Coyote *(Canis latrans)*, Gray Fox *(Urocyon cineroargentus)*, or, worse, a Western Meadowlark *(Sturnella neglecta)*, a bird feared for its harmful power more than any other animal.

Owls were some of the few animals not eaten, for religious reasons, at least by the Ohlones (Margolin 1978) and probably many other tribes, but owl feathers were made into ceremonial cloaks and blankets and appeared in headdresses (Heizer and Elsasser 1980). The feathers of the Great Horned Owl seemed to be particularly favored, perhaps because of their large size.

The personal names of Miwoks (central California) were often evocative and extremely specific, based on very careful observations of their natural surroundings. "Bobbing like a topknot of a walking quail," was one surely charming girl's name, and the male name Tiponya translates to "Great Horned Owl sticking head under its body and poking egg when hatching" (Bibby 2005, 121–122). The names of the owls themselves were often onomatopoeic versions of their calls (as are some bird names in English); for example, "chah-he," a name used by San Juan Bautista Costanoans, is a reasonable approximation of a Barn Owl's call.

Owls in the Modern World

In the twentieth century, it was common practice among European (and to some extent North American) hunters to tether a large owl or mount a stuffed one in an open place by a blind and shoot the hawks, falcons, and other raptors and corvids that cannot resist harassing the feared predator. With the exception of certain Mediterranean countries, this practice has largely stopped, but not the killing, in some places. In Malta, for example, where for centuries the attitude has been "if it flies, it dies," it is now illegal to shoot raptors, but migrating Short-eared Owls *(Asio flammeus)* turn up in rehab centers with gunshot wounds, and thousands of other birds of prey are killed each year.

Stuffed or artificial owls are sometimes still used in America to attract flying raptors within "shooting" range of photographers, and rehabilitated but unreleasable Great Horned Owls, legally possessed by biologists under permit, serve to draw nesting raptors (among them hawks, falcons, and other owls) into nets for banding and radio tagging and for making scientific measurements (Bloom et al. 1992). A less satisfactory enterprise, at least from the consumer's end, is the brisk sale of plastic Great Horned Owls, which are mounted on roofs for the purpose of driving away nuisance feral Rock Pigeons *(Columba livia)*. These "scarepigeons" work only very briefly, if at all, and are in fact soon sought out by the pigeons because they provide a modicum of shade on a sun-drenched roof.

Fig. 80. A plastic Great Horned Owl, mounted on a roof to drive away pigeons, is often regarded by the latter as a friend or a source of shade.

Fig. 81. The Great Horned Owl is the only owl species in the United States that is (and is legally allowed to be) used in falconry.

Although a bit dull during the day, owls, especially the small species, make interesting pets, and many a biologist raised one as a youth in the days when it was legal to do so. All are now, however, protected by law, and it is illegal to keep them with the exception of the Great Horned Owl, the use of which, by license, is permitted for wildlife study purposes and in falconry, the sport in which birds of prey are trained to hunt game at the bidding of humans. Even in this endeavor, owls are less than ideal; although a powerful raptor like a Great Horned Owl can readily catch rabbits, one individual intended for that quarry was inordinately fond of Western Toads *(Bufo boreas)*, to the displeasure of the owl's keeper. Besides, trained owls are much more interested in hunting after dark, which of course makes it impossible to enjoy the flight, and should the falconers bring along lights to illuminate the scene, they are likely to be quickly importuned by alarmed citizens and armed authorities.

An easier and more popular pastime is collecting owl art.

Giant owls carved out of redwood are sold at tourist stops all along the northwest coast, and elegant gift shops everywhere sell porcelain and glass miniatures eagerly collected by some. Owl depictions, too, especially in the form of limited edition prints, have their loyal following; the Snowy Owl is a favored subject.

Owls are also popular commercial symbols and not uncommonly turn up in corporate logos. But no owl has entered the services of humans as extensively as the Barn Owl.

The Barn Owl Industries

Because of their extraordinary talents of navigating silently and finding food in total darkness, combined with their ease of maintenance in captivity, Barn Owls have become subjects of research in a dazzling diversity of scientific fields.

Much research has been done on the orienting behavior (visual and auditory mapping; see the section "An Owl's Body") of Barn Owls, resulting in some interesting applications. Stroke victims and people with brain injuries may be helped by studies demonstrating that older Barn Owls can relearn skills (adjust their visual and auditory maps) substantially better if they are taught in incremental, step-by-step fashion (Linkenhoker and Knudsen 2002). Another study links Barn Owl neuroscience with robotics: a robotic assemblage, controlled by a model of the principal neural structures of a Barn Owl's brain (again, learning), helped researchers design a system possessing the flexibility needed to function autonomously in an environment with continually changing conditions (in other words, the real world) (Rucci et al. 1999). Someone even built an analog silicon chip using the algorithm of the Barn Owl neural circuitry, reproducing the steps through which the owl measures interaural time differences, creating a chip that used less energy and less space.

In another direction, the noise-reducing modifications of the owl's wings are of great interest to researchers investigating noise reduction of landing airplanes (apart from engine sound), which would make possible increased traffic at major airports, an important economic consideration to airlines (Lilley 1998) and to designers of military aircraft (Thomas et al. 2002).

There are less technical Barn Owl industries as well. The Barn Owl's usefulness for rodent control has been more fully appreciated in recent years. Installation of Barn Owl nest boxes in oil

Fig. 82. The thick pile on a Great Horned Owl's primary resembles velvet. Barn Owl remiges are similarly endowed. (See fig. 27.)

palm plantations in Malaysia resulted in a decided decline of destructive rats and much reduced use of rodenticides (Bruce 1999); however, in Hawaii, which had no Barn Owls before some were brought from California for rat control, the raptors negatively affect some indigenous seabird populations (U.S. Fish and Wildlife Service 2005).

More recently, nest boxes for this species have proliferated through much of California (and elsewhere), particularly in the orchards of the Central Valley and in vineyards. Many thousands of Barn Owl boxes—along with similar numbers of homes for Wood Ducks *(Aix sponsa)*, American Kestrels *(Falco sparverius)* (both of them sometimes also used by owls), and Western Bluebirds *(Sialia mexicana)*—were built by a Merced woodshop teacher and his students (S. Simmons, pers. comm. 2005). Research in 1997 and 1998, using his boxes, showed that Botta Pocket Gophers *(Thomomys bottae)* constituted 73 percent of the owls' diet by weight, with California Meadow Voles *(Microtus*

Fig. 83. Barn Owl boxes mounted on poles in vineyards have become a common sight in wine country.

californicus) contributing another 20 percent, and deer mice *(Peromyscus)* 4 percent—all rodents harmful to agriculture. Simple math revealed that 48 pairs of Barn Owls, each with three surviving owlets (actually, more often five to seven), would consume over 6 tons of Pocket Gophers per year (S. Simmons, pers. comm. 2005)! A proliferation of Barn Owls resulting from the provision of new housing is, however, less than desirable in areas that are home to endangered rodents, as is the case in parts of California, where several species or subspecies of kangaroo rats are endangered and barely hanging on, such as the Giant Kangaroo Rat *(Dipodomys ingens)*. Artificially increasing such efficient predators as Barn Owls could easily push these beautiful animals over the edge (P. Bloom, pers. comm. 2007).

Apart from lively nest box sales, a small cottage industry fur-

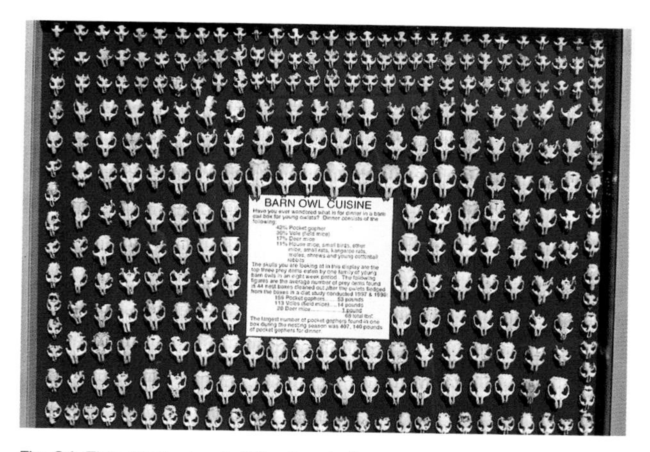

Fig. 84. This display box holding the skulls and skull fragments of mainly California Meadow Voles and Botta Pocket Gophers demonstrates the numbers of rodents consumed by one brood of Barn Owls during an eight-week period.

nishes owl pellets (mainly those of the Barn Owl) for biological supply companies, which in turn sell them to schools and educational nature centers for dissection by young students. Because it is possible to contract a salmonella infection from pellets, they are typically sterilized. For a simple assessment of contents, pellets may be broken up by hand, perhaps with the aid of forceps and a dissecting needle.

A more detailed analysis, especially of large samples, is facilitated by soaking the pellets in an ample solution of sodium hydroxide (100 g NaOH crystals per liter of water) for two to four hours and then washing the soup through a quarter-inch mesh screen to catch the bones and other fragments that can be used to identify and quantify the prey (Marti 1987). Aluminum leg bands, which are affixed to migrant birds to track their movements, are sometimes ingested by owls along with the leg and must be removed before soaking, because the NaOH destroys them and the information they bear (Schueler 1972). Modern paleontologists study fossilized owl pellets, which contribute important information to the Tertiary small-vertebrate fossil record.

Owl Conservation

Owls in Trouble

Every year, hundreds of Barn Owls die in California from a variety of causes; yet, this species is holding its own in most of the West and certainly in California, in contrast to parts of Europe and some other parts of North America, where the bird is in decline. It is ironic that a species that has so long associated itself with humans should find itself in difficulties because of them.

A few owl species have actually benefited from human activities, at least locally; for example, buildings provide nest sites for the Barn Owl, and some owls take advantage of street lighting to catch invertebrates attracted at night. Three species of owls (the Barn Owl, the Western Screech-Owl *[Megascops kennicottii]*, and the Great Horned Owl) have been observed downtown in the California city of Walnut Creek (Contra Costa County), indicating that this rather unorthodox habitat can provide food for these predators. Some owls, however, are in trouble in California and elsewhere in the West, mainly because of habitat destruction.

The conservation issues facing owls in California are presented in this section by way of introducing the major environmental challenges (along with their inadventent or collateral problems) that owls encounter throughout their ranges, and the creativity and science needed to meet them. For additional conservation and status information on owls in other western states, see the species accounts.

Only one California owl has been listed under the federal Endangered Species Act. Two others are protected under the California Endangered Species Act, and the rest are designated as California species of special concern, which, however, gives the species no special legal status; such designation is intended for use as a management tool only.

Northern Spotted Owl (U.S. Threatened) California Spotted Owl (California Species of Special Concern)

The Northern Spotted Owl *(Strix occidentalis caurina)*, a subspecies of the more widespread Spotted Owl, occurs in a variety of chiefly coniferous habitats, one of them being the old-growth coastal forests of the Pacific Northwest (including those of northwest California), where this bird has become famous as the icon of those ancient woods, though it is by no means restricted to this coastal zone. It is estimated that 80 percent of these forests have been logged off over the past 200 years, causing habitat loss for the owl as well as habitat fragmentation. The bird's population in this area is declining, and many other species are affected by the felling of the great trees, which can take centuries to grow.

Leading up to the Northern Spotted Owl's federal listing as threatened in 1990 was ever increasing acrimony between conservationists and the timber industry. Bumper stickers reading "Save a logger, eat an owl" became common in logging areas, and local restaurants facetiously offered owl barbeques. In the summer of 1989, the U.S. Forest Service decided not to send two of its regular participants, Smokey Bear and Woodsy the Owl, to a yearly Oregon parade, reportedly because it had received death threats against the two (Yaffee 1994). On the other side, activists sabotaged tree trunks with metal or ceramic spikes that would cause saw blades in sawmills to shatter and inflict injury on the sawyers (Guynup and Ruggia 2004).

Not long ago, experts agreed that survival of the Northern Spotted Owl required saving blocks of suitable habitat, that is, old-growth forest areas large enough for "all life functions for multiple pairs of Northern Spotted Owls which should be distributed throughout the current range of the subspecies" consistent with dispersal distances of juveniles and "embedded in a landscape of habitat that allows foraging, resting, and cover by birds dispersing between blocks" (Verner 1992, 302). The preservation of such an enormous area obviously runs counter to economic interests and currently prevailing political philosophies.

It comes as some consolation that the owl lives in other habitats and can also survive in second-growth forest that is at least 30 to 40 years old. A forest that is managed (that is, from which trees are removed) can still sustain owls if it contains elements of old, unmanaged forests: very large trees, structurally diverse canopies, shade-tolerant conifers, and large snags and fallen trees (Horton 1996). The owls require stands of old trees for roosting and nesting (Folliard et al. 2000), and they selectively seek out remaining patches of a more complex forest structure in second-growth areas (Gutiérrez et al. 1998). But coastal Northern Spotted Owls are more numerous and inclined to nest in old forests (Guitiérrez et al. 1995).

Although the Clinton administration in 1994 preserved millions of acres for Spotted Owl protection through its Northwest Forest Plan, the gains made by that action were substantially rolled back by the Bush administration through placing roadless areas under state purview, which exposes them to the pressures of economic interests. The Bush administration has also promoted increased logging on federal lands under the guise of fire protection, when in fact the brush that replaces old-growth is far more flammable than the big trees. Ninety percent of old-growth forests are on federal land.

The U.S. Fish and Wildlife Service announced that its recovery plan for the owl will be completed by 2008; it will address owls on private land as well as federal land. Conversion of private forest lands to vineyards, such as in Napa County, also may take habitat from these owls. Landowners must apply for an "incidental take" permit, required under the Endangered Species Act, and agree to some mitigation measures (for example, maintaining a portion of the land as owl habitat).

The dominant threat to this owl's survival, however, is the invasion of the Barred Owl *(Strix varia)* from the north, likely the result of human-caused alterations of the landscape. First documented in California in 1981 (Evens and LeValley 1982), it has reached Marin County and is scattered through the Sierra Nevada down to the southern tip. The Barred Owl endangers the continued existence of the Northern Spotted Owl in several ways. It sometimes eats its slightly smaller cousin, and it displaces it in some areas; it even hybridizes with it (Hamer 1993). The hybrids resemble the Northern Spotted Owl sufficiently, and at least some of them are fertile, so that they may (and do) backcross with pure Northern Spotted Owls, leading to dilution of the bird's gene pool. However, a study in the central Cascades of Washington suggests that where Barred Owls have become firmly established in Spotted Owl country, no hybridization occurs (Herter and Hicks 2000), the phenomenon thus being associated with the disturbed conditions at the forefront of the Barreds' invasion.

In Washington State, where the owl arrived earlier (1973), Spotted Owl populations are down to one-half or one-quarter of their original size, surviving chiefly at higher elevations, with the Barred Owl preferring flatter ground, alluvial habitat, and gentler slopes (Buchanan et al. 2004); nevertheless, it is also moving upslope. In California's Redwood National Park (Humboldt County), Barred Owls have completely replaced Spotted Owls

(B. Woodbridge, pers. comm. 2006). Because the Barred Owl has smaller home ranges than the Spotted Owl, a given area can support a larger population of the invader.

In 2005, two pairs of Barred Owls in the Klamath National Forest were shot for the scientific collection at the California Academy of Sciences in San Francisco. Spotted Owls promptly moved back into the suddenly vacant territories (B. Woodbridge, pers. comm. 2006)—whether to breed remains to be seen. It is difficult to see how such "cleansing" would work on a larger scale and over long periods of time, given the continued influx of Barred Owls. Sterilization of female Barred Owls is among other contemplated control measures.

The Great Horned Owl, too, takes advantage of altered landscapes, moving into clear-cut areas and using the new forest edges (Ehrlich et al. 1992). Although it tends to occupy more fragmented forests than the Spotted Owl, in areas occupied by both species, the Great Horned is a competitor and predator of the Spotted, particularly juveniles (Gutiérrez et al. 1995).

Only the Northern Spotted Owl is federally listed as threatened in California, as well as Oregon and Washington, but this state is also home to a closely related subspecies, the California Spotted Owl *(Strix occidentalis occidentalis)*, which occurs in the Sierra Nevada and in other mountain ranges in southern California and nests in large old oaks *(Quercus)* as well as conifers. These owls, too, are affected by the invader and by clear-cut logging and development, including the rapid urbanization of the Sierra foothills, where many Sierra owls spend the winter. Individual populations in southern California have the potential to become isolated (Steinhart 1990). Petitions to list California populations of the California Spotted Owl federally have been denied, most recently in 2006.

Elf Owl (California Endangered)

In the early 1900s, a small population of Elf Owl *(Micrathene whitneyi)* was found on the California side of the lower Colorado River, breeding in saguaros *(Cereus giganteus)* in an area that is now under Squaw Lake. Poor water management practices (including the damming of the river and subsequent agricultural development along it) severely damaged the owl's habitat. Saguaros have been essentially extirpated from California, and the native trees such as cottonwood (*Populus*), paloverde *(Cercidium)*, mesquite *(Prosopis)*, and willow *(Salix)*, which provide holes (usually excavated by woodpeckers) for nesting, are disappearing. Tamarisk (salt cedar *[Tamarix]*), an invasive nonnative tree, lowers the water table because of its deep roots, and it is displacing the native trees at an alarming rate. Unlike native desert trees, tamarisks lack cavities that might serve as nest sites for Elf Owls (P. Bloom, pers. comm. 2005).

The species was listed by the state as endangered in 1980. A revegetation project begun in 1983 had mixed success, and owl reintroduction efforts west of Needles, California, in 1986 were unsuccessful, apparently because of the degraded habitat (Henry and Gehlbach 1999). In 1987 there were probably no more than 25 pairs of Elf Owls breeding in California (Holt et al. 1999).

The Elf Owl may no longer be a California species. Searches in 1999 in Paiute Canyon, a reintroduction site, were fruitless (G. Gould, pers. comm. 2005).

Great Gray Owl (California Endangered)

The Great Gray Owl *(Strix nebulosa)* suffers from exceedingly low population numbers in California, although it is more common in other states in its range. Before logging reduced the forests (where they breed) and sheep and cattle grazing damaged many mountain meadows (where they forage), this species was probably more abundant and widespread, even nesting in northwestern counties of the state (Steinhart 1990). Only 17 nests have been found in California (Beck and Winter 2000). California listed the species as endangered in 1980.

The forests in and around Yosemite National Park, which have been spared from logging and grazing for over a century, are where most known nesting occurs; however, the development of campgrounds in the park probably eliminated some of the owl's foraging areas, and, considering the number of visitors to the park each year, human recreational activities may disturb owls at the nest. Collisions with cars in the park killed 16 Great Gray Owls from 1966 through 2005.

Outside of the park boundaries, timber harvesting may be the biggest threat to this species, followed closely by grazing. This owl is highly dependent on meadows for its survival in California, partly because Sierra forests generally lack other grass-forb habitats (Beck and Winter 2000). Its requirements are rather specific: meadows, for example, must be large enough and have enough cover by the end of summer to hold a good supply of rodents (Beck and Winter 2000). Strychnine poisoning of pocket gophers *(Thomomys)* may harm these birds (Bull and Duncan 1993), as might prescribed burning.

Forest management practices designed to maintain habitat for the Great Gray Owl may include leaving sufficient large-diameter-topped trees (needed by the owls for nesting) as well as leaning and fallen trees (climbable roosts for young before they can fly) and dense canopies (hiding spots for juveniles). In logged areas, some perch trees should be left so that the owls can hunt; some cutting can actually enhance the owl's habitat by opening up dense forest (Bull and Duncan 1993).

Human-made nest platforms have been readily accepted by these owls elsewhere, and their use appears to result in a higher nesting success rate than that of natural sites (Bull and Henjum 1990) and could also potentially allow this species to nest in new areas.

The Great Gray additionally shares its habitat with the Northern Goshawk *(Accipiter gentilis),* a powerful predator, and the equally potent Great Horned Owl. Both can have a major impact on a population that is dangerously small to begin with (B. Woodbridge, pers. comm. 2006).

Burrowing Owl (California Species of Special Concern)

This owl is generally declining in distribution and abundance throughout its range in North America (Haug et al. 1993). It is listed as a species of special concern in many western U.S. states, including California, which supports the largest numbers of both resident and wintering owls (Sheffield 1997). It is listed as endangered in Canada and threatened in Mexico, state-listed as endangered in Minnesota and threatened in Colorado, but it has no federal regulatory designation in the United States.

The Burrowing Owl was once widespread and locally abundant in California; a decline in populations was first noticed in the 1940s (Grinnell and Miller 1944). A survey conducted in the early 1990s revealed that in the preceding decade alone, over half of the state's known breeding colonies had disappeared (although additional colonies were discovered), the number of breeding pairs per breeding colony had decreased (especially in the Central Valley), and the owl had been extirpated or nearly extirpated as

Fig. 85. All too often, Burrowing Owls are reduced to perching on construction stakes as yet another field becomes a parking lot.

a breeding species from nearly a quarter of its range, that is, from most coastal counties as well as from the northern ends of San Francisco, San Pablo, and Suisun bays (DeSante et al. 1997).

The major cause of the decline is most likely habitat destruction, chiefly by the paving over of grassland for commercial and residential development and highways. Even the well-meaning restoration of salt marshes by flooding grasslands adjacent to San Francisco Bay has taken away vital habitat (Johnson 2004). Buffer zones would help Burrowing Owls (San Francisco Estuary Institute 1999). These birds adapt to human-modified landscapes, provided there are food and nest sites in suitable habitat; a transitional grassland strip between original or restored high marsh and lands converted to human use suffices as habitat (Trulio 2004).

In the city of Fremont (Alameda County), a pair of Burrowing Owls set up housekeeping in an old Beechey Ground Squirrel *(Spermophilus beecheyi)* hole located in a small triangular island of bark chips next to a parking lot, between lawn-covered sports fields and a weedy field, which provided food for the family. In fact, a study on the New Mexico State University campus showed that Burrowing Owls nesting there had significantly more nestlings and fledged more young than pairs nesting in natural habitats, likely because of access to much more food and safety from predators (Botelho and Arrowood 1996).

Because this owl usually depends on the holes of fossorial (burrowing) mammals for nesting, the elimination of ground squirrels (as, for example, during the planting of vineyards), along with habitat fragmentation, shooting, and contamination with pesticides, have all played a part in the owl's precipitous decline (Haug et al. 1993). Urbanization no doubt leads to an increased mortality rate from collision with vehicles where the owls nest near roads, and it can also increase predation pressures: Coyotes *(Canis latrans)* (whose populations are artificially enhanced by development) and dogs not only kill owls but also destroy burrows (Zeiner et al. 1990).

On the Carrizo Plain, predation by raptors was the leading cause of Burrowing Owl mortality (Klute et al. 2003). On Santa Barbara Island (one of the Channel Islands), a population of about 20 Burrowing Owls was extirpated by Barn Owls in 1984 after the cyclic decline of the Barn Owl's rodent prey (Drost and McCluskey 1992). Clearly, some losses are unavoidable.

The Burrowing Owl coexists with agriculture in the Imperial Valley, southern Central Valley, and lower Colorado River Valley. In fact, the Imperial Valley, which historically had few owls before the 1900s (before the introduction of industrial agriculture), now has one of the highest densities of Burrowing Owls in its entire range (DeSante et al. 2004). Seventy-one percent of California's Burrowing Owl population lives here, on 2 percent of its range (J. Barclay, pers. comm. 2007). The owls can be seen in numbers in this almost entirely human-made environment of levees (with squirrel, gopher, Muskrat *[Ondatra zibethicus]*, and erosion holes) and agricultural fields, often alfalfa *(Medicago sativa)* and Sudan grass *(Sorghum bicolor)*, crops that are popular with rodents. But life for owls in the Imperial Valley is not risk free. Dredging of agricultural drains has destroyed nests. Adults and nests have been buried by road grading, and flooding from overflowing agricultural ditches has also caused nest destruction and the death of young (Rosenberg and Haley 2004). More than 300 pairs have been lost to habitat destruction from 2001 to 2006 (P. Bloom, pers. comm. 2006).

In 2003 the California Fish and Game Commission rejected a petition to upgrade the state listing of the Western Burrowing Owl *(Athene cunicularia hypugea)* to threatened, finding that

Fig. 86. A Burrowing Owl on a berm in the Imperial Valley (Imperial County) assuming a stance similar to the "white and tall" position used in courtship.

although habitat loss caused by urban development is an immediate and serious threat to survival of breeding populations in coastal areas, such local extinction does not constitute a "significant portion" of the species' range; therefore, the owl did not require increased government protection.

Obviously, because over 90 percent of California's owls live on private land (DeSante and Ruhlen 1995), the survival of the Burrowing Owl is largely dependent on the attitude and management practices of private landowners and irrigation districts.

The Burrowing Owl is protected by both state and federal laws. For example, the federal Migratory Bird Treaty of 1918, amended in 1972 to add eagles, hawks, and owls to the list of protected species, makes it illegal to take (a term that includes destroying nests or killing the birds), possess, buy, sell, or barter them or parts thereof. However, the law is not protective enough to stop the slaughter of hundreds of Burrowing Owls each year by wind turbines at the Altamont (Alameda County), along with great numbers of other raptors and nonraptorial birds. The California Environmental Quality Act stipulates that significant impacts to Burrowing Owls be mitigated; in other words, the destruction of habitat is allowed except during the nesting season, although compensation may be required (Trulio 1997). Some landowners, fearful that Burrowing Owls may be discovered on their property, disk their land to prevent owls from using it.

The California Burrowing Owl Consortium developed a Burrowing Owl Survey Protocol and Mitigation Guidelines to help evaluate and ameliorate impacts from development projects in the state (access at www2.ucsc.edu/scpbrg/section1.htm). These guidelines need some revision; for example, because foraging distances by nesting males was highly variable, but chiefly within 600 m (1,968 ft) and as far as 2 to 3 km (1.2 to 1.8 mi) in one study (Rosenberg and Haley 2004), the suggested allotment of 2.6 ha (6.5 acres) per pair of nesting owls, calculated on a 100-m (328-ft) radius around the burrow, is inadequate.

Long-eared Owl (California Species of Special Concern)

Eighty years ago, the Long-eared Owl *(Asio otus)* was a common breeding bird in California, even abundant in some areas. The Sacramento and San Joaquin valleys and the San Diego area were all identified as "centers of abundance," along with the northeastern Great Basin area. But today, breeders are scarce, and this owl is probably extirpated in the Sacramento Valley, and in the Escondido area (San Diego County), the species has not been seen in years.

Although the destruction and fragmentation of riparian woodlands, oak groves, and grasslands are thought to be the chief reason for the Long-eared Owl's decrease, the owl was in fact reported to be in decline before these habitats were reduced (Grinnell and Miller 1944). Some populations may have been affected by collisions with cars, to which this species is very susceptible (California Department of Fish and Game 2003). Predation by Red-shouldered Hawks *(Buteo lineatus)* and Red-tailed Hawks *(B. jamaicensis)* was documented in southern California. The Common Raven *(Corvus corax)*, steadily increasing in numbers, preys on this owl as well but at the same time also provides additional nest sites for it (Bloom 1994).

Short-eared Owl (California Species of Special Concern)

A decline in wintering Short-ears in California was noted decades ago (Grinnell and Miller 1944). In May 1936, large numbers, evidently breeders, were observed near Lava Beds National Monument and Tule Lake, and within one hour, 12 individuals came to drink at a drainage ditch (Dixon and Bond 1937); however, the owl's population and range have since steadily decreased.

Because of this species' nomadic habits, it is difficult to arrive at accurate numerical changes. Estimated population declines of from 69 to over 80 percent in North America since the 1960s have been proposed. The owl is in peril in most of its range, especially the prairie provinces of Canada, along the Pacific Coast, and parts of the southeast and northeastern United States (Ehrlich et al. 1992). Some northeastern states have listed it as threatened.

As with other owls, loss of habitat is probably largely responsible for this bird's decline in California. In the Sierras, most suitable habitat of this species has been destroyed by sheep grazing, water diversion, and recreational development. Likewise, much habitat in the Central Valley has been greatly reduced by cultivation and the draining of marshes (Verner and Boss 1980). Coastal and inland grasslands (major habitats for this species) lend themselves all too easily to urban development, and marshlands are drained for agriculture or filled for business parks and housing.

Unfortunately, Short-eared Owls are easy to shoot and make tempting targets. Predation likely played a substantial role as well; as a ground nester, the species is a frequent victim of Striped Skunks *(Mephitis mephitis)*, which eat eggs and small young, as do Red Foxes *(Vulpes vulpes)* and Coyotes, and its nests are destroyed by mowing machines in hayfields. The increase in American Crows *(Corvus brachyrhynchos)* in some parts of California in recent decades also may have had an impact on nesting success, because they eat the eggs of this and other raptors. Great Horned Owls eat young and adults, as do certain diurnal raptors.

Hazards to All Owls

The no longer deniable change in climate may not, at first glance, seem obviously problematic for owls, but its consequences are far reaching. In the Sierra Nevada, certain mice are now found at much higher elevations than they were 60 years ago (J. Patton, pers. comm. 2005), raising the possibility of range expansions of owls that prey on them, which might bring them in conflict with species confined originally to higher elevations. Another unknown is how forest health will be affected by global warming and, along with it, the prey base (such as insects, the near-exclusive diet of Flammulated Owls *[Otus flammeolus]*).

Cyclic populations of voles in Scandinavia have, in recent decades, undergone a dramatic decline because of the loss of snow cover resulting from increasingly mild winters produced by a change in the North Atlantic Oscillation. This periodic climate change, the cause of which is unknown, has indirectly affected, by way of the voles' abundance, the populations of Tengmalm's Owl (called the Boreal Owl *[Aegolius funereus]* in North America), and both prey and owls now demonstrate minor, more or less annual fluctuations instead of the dramatic high-amplitude cycles of bygone years (Hörnfeldt et al. 2005).

An important threat to all North American owls is West Nile Virus, a mosquito-vectored disease that is particularly pernicious to owls, among other bird groups. Little is known about the effects of this virus on California populations so far because it has just recently spread to this state, but it has devastated owl populations elsewhere in the country. Although some years show fewer avian mortalities than others, the virus is not going to go away, and bird (and a few human) casualties can be expected indefinitely. Local health department and vector control agents test dead birds for the virus; if you find one that is fresh and appears to have died from the disease (not trauma), in California call 877-WNV-BIRDS.

Nearly all owls in the West, and especially California, suffer to some degree from human-linked dangers, some more so than others. The destruction of habitat extends from the coastal coniferous forests to the interior oak woodlands, grasslands, and riparian woodlands, and the deserts as well. In Alameda County, an ephemeral stream that was bulldozed out of existence prior to the construction of office buildings took with it a modest willow riparian woodland, the roost and nest site of a family of Barn Owls. Following the work, the owls perched forlornly on the remaining low berms, without any cover. In southwestern California, breeding Long-eared Owls, unable to adapt to urban pressures, declined by 55 percent (Bloom 1994). Salvage logging and the removal of dead trees for fire suppression reduces the number of potential nest sites for hole-nesting species.

Owls are all too common victims of motor vehicles, none more so than the Barn Owl, whose foraging flight altitude equals the height of a semi. Western Screech-Owls and Great Horned Owls, too, can be found dead on roads and highways with distressing frequency, and even maimed Northern Pygmy-Owls *(Glaucidium gnoma)* are brought to rehab centers.

Pesticides also take their toll, chiefly through secondary poi-

Fig. 87. Broken wings blow forlornly in the slipstream of cars as yet another Barn Owl falls victim to vehicular traffic.

soning, when owls eat prey that has been killed or disabled by these insidious toxins. Carbofuran, a particularly nasty insecticide, has been shown to have killed Short-eared, Burrowing, and Great Horned Owls, in addition to a great variety of diurnal raptors. Directions for use explicitly state that this poison is not to be applied around birds and other wildlife; users apparently assume that such are not found in agricultural fields (Mineau et al. 1999). Rodenticidal anticoagulants (which cause the consumer to bleed to death internally) killed a variety of owls that had eaten poisoned rodents (Blus 1996). Chlorophacinone and diphacinone, used for ground squirrel control in central California, were implicated in the deaths of Great Horned Owls and a Barn Owl (Peeters 1994).

The shooting of owls is unfortunately not a thing of the past, though it is much less common than it was 30 and more years ago. Owls become entangled in wire fencing as they pursue a prey animal on the opposite side, and they injure themselves on barbed wire. Prospecting for nest sites, they sometimes enter pipes from which they cannot escape. Wind turbines, especially those at the

Fig. 88. Wire fences can prove deadly to hunting owls such as this Great Horned Owl, who perhaps regard these structures as weeds that yield.

Altamont (Alameda County), take a terrible toll not only of Burrowing Owls but also of other species.

A study of Spotted Owls nesting in the Canyonlands and Capitol Reef National Parks in Utah revealed that female owls were sufficiently disturbed by hikers that they spent nearly 60 percent less time on activities such as feeding their young and a third less time on daytime maintenance activities such as preening their young and themselves (Swarthout and Steidl 2003). Hikers staying about 25 m (80 ft) away resulted in 95 percent of the birds not taking flight. Obviously, buffer zones around known nests should be mandatory.

Bird watchers eager to see owls can have a profound impact

on the success of nesting by using tape-recorded songs to draw the birds into visual range. In the Corkscrew Swamp of Florida, the Barred Owl capital of the United States, not one pair bred successfully one year because of overuse of tape recorders (de la Torre 1990). In a long-term Arizona study of 24 Whiskered Screech-Owl *(Megascops trichopsis)* nests, 14 visited by birders and 10 left alone, the disturbed ones showed effects ranging from the loss of one egg to total clutch failure to undersize fledglings (Hanson 2000). See also "Direct Evidence" in the section "Finding and Watching Owls."

Allowing owls to reproduce in peace would seem an obvious measure to limit nest failure. It is bad etiquette to play tapes, imitate songs, or touch nest trees during the breeding season. Raccoons *(Procyon lotor)* are known to investigate trees with human scent and would not hesitate to help themselves to eggs or young where accessible. A conservation-minded birder would follow some rules: males should be located only at their day roosts, and any contact with owls should be avoided during the first hour after dark, the most important feeding period during the breeding season; this means you should set out for an owling expedition after that first hour (Hanson 2000).

Living with Owls

Given a chance, many owls do not seem to mind at all living around humans. Burrowing Owls adapt well to the human-made landscape, and Barn Owls are famous for their association with buildings and agricultural fields. Great Horned Owls find homes in suburban stands of eucalypts and in groves on college campuses. A New Jersey study showed that Barred Owls avoided suburban habitat, but Great Horned Owls had a tendency to move in (Bosakowski and Smith 1997). Western Screech-Owls can be found in Berkeley (Alameda County), for example, and in suburbs, parks, and gardens of other cities. A study of the closely related Eastern Screech-Owl *(Megascops asio)* has shown that suburban life enhances egg and chick survival, making these birds more productive than their country kin, whose nesting success was half that of the suburbanites. Probably the chief benefit is the reduced presence of competitors and of predators that overwhelm the rural ones. The urban populations are more stable than the rural ones, with improved adult survival, because of the benefits of higher prey densities, excellent hunting habitat, and climatic stability (Gehlbach 1996).

Where landowners have the foresight of preserving snags and cavity-holding trees on their property, they may find even more elusive species nesting where people can see and enjoy them. And when there are no snags, nest boxes are often very successful substitutes.

Nest Boxes

It becomes evident just how strapped some cavity-nesting owls are for nest sites when nest boxes are placed in vineyards and orchards. In late winter these new homes are typically found and occupied within a matter of weeks by seemingly homeless Barn Owls. Western Screech-Owls and Northern Saw-whet Owls *(Aegolius acadicus)*, too, readily accept boxes (often very close to houses), notwithstanding the apparent abundance of large woodpecker holes in their habitats. This, however, does not mean that every box put up will necessarily be occupied. If not taken after two nesting seasons, it should be moved to a new site. For optimal success, boxes should be put up well before the onset of breeding, which is January and February. Insecticides and rodent poisons should not be used in the area, and tree trimming should be done in fall. It must be borne in mind that owls generally do not hunt in the immediate vicinity of their nest box but will allow foraging in their yard by neighbors.

With fine housing provided for them, unexpected (and sometimes unwanted) tenants may move in. If the squatters are American Kestrels, most landlords would be delighted. If they are squirrels *(Sciurus)* (which often chew up the entrances) or European Starlings *(Sturnus vulgaris)*, which make scarce nest sites unavailable to native species, eviction is desirable.

Many Web sites offer nest box pointers and designs, but some plans are better than others. Most guidelines, however, are pretty standard, although everyone seems to have a different idea about the best size for the entrance hole. Above all, a good box needs to protect the occupants from rain, excessive heat, and overcrowding.

Use any lumber that is suitable for outdoors; exterior plywood

BUILDING A NEST BOX: SOME TIPS

CLIMATE CONTROL. Adequate ventilation must be provided by drilling half-inch holes in the bottom (which also help to keep the floor dry) and under the roof overhang (or leaving a gap under the roof). Overhanging foliage helps in keeping the box cool. Owls may abandon a nest if the box gets too hot.

COZY AND CLEAN. A one-inch layer of wood shavings on the floor will prevent the eggs from rolling about and facilitate cleaning the box after the owlets have fledged. A flat roof without an overhang (or one that is divided so that only part of the roof can be lifted up) might allow rain to enter.

FALL CLEANING. A large hinged port on the lower half of one side of a Barn Owl box allows for annual cleaning; a box side that swivels on pins to open also works well on boxes for small owls. Cleaning is best carried out in October and November, after which the box can be rinsed with a 10 percent chlorine bleach solution or with boiling water to kill parasites. The use of a dust mask and latex gloves is recommended.

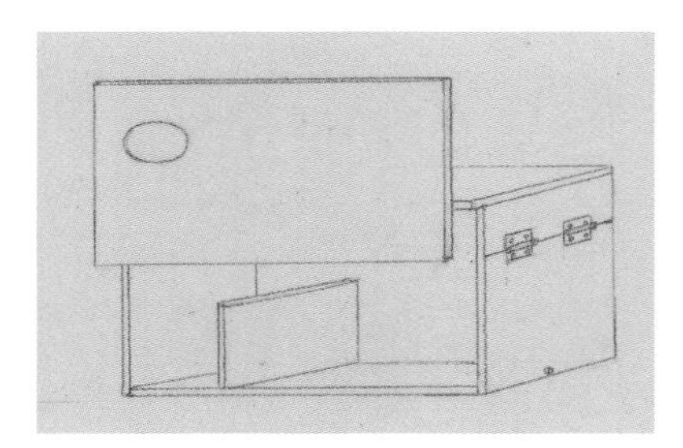

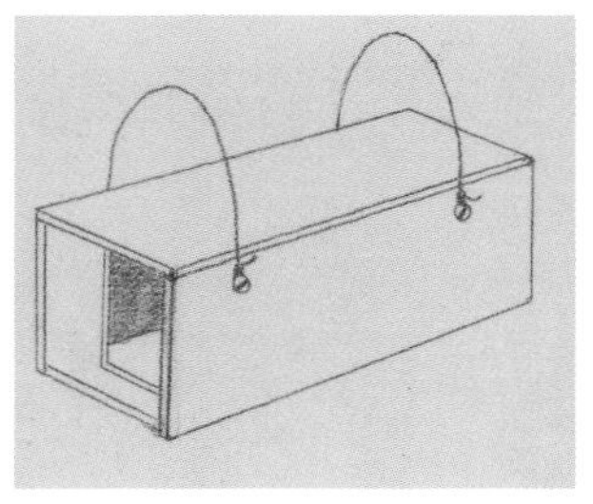

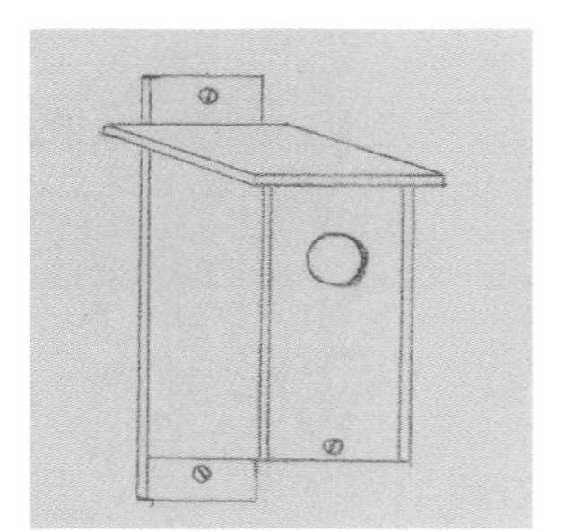

Fig. 89. Nest box diagrams *(clockwise from top left):* Barn Owl (note baffle to keep raccoons from reaching in to rob the nest); Barn Owl ("Bloom Box"); screech-owl or saw-whet owl.

Species of Owl	Inside Floor Dimensions	Entrance Hole Diameter or Size	Center of Entrance Hole Above Floor	Front Panel Height (see note below)
Northern Saw-whet	7 in. × 7 in.	2 in.	13 in.	16 in.
Western Screech	8 in. × 8 in.	3 in.	12.5 in.	16.75 in.
Barn	12 in. × 30 in. (minimum; includes 7.5 in. × 12 in. vestibule)	3.75 in. × 4.5 in. (horizontal ellipse)	8.5 in.	16.75 in.
Barn (Hanging "Bloom Box")	14 in. × 30 in.	7 in. × 12 in.	NA	(Back Panel) 13.75 in.

Sources: Northern Saw-whet from Point Reyes Bird Observatory, Stinson Beach, California (slightly modified); Western Screech from Kaufman (2002) (slightly modified); Barn from S. Simmons, Raptor Works, Merced, California; Barn (hanging box) from P. Bloom.

ROOM WITH A VIEW. Suggestions as to mounting heights for boxes vary. Most indicate 12 ft or higher for Barn Owls and 10 ft or higher for the smaller species. Barn Owl boxes mounted a mere 8 ft from the ground to the roof of the box have been very successful in the Central Valley, however, and the smaller owls do not seem to object to mounting heights of 9 ft or even less.

BUILT TO SPECIFICATION. Here are the perhaps ideal dimensions for four types of owl nest boxes. Whichever you build, be sure to do the following:

- Add 1.5 in. to the rear wall of screech-owl and Northern Saw-whet boxes to obtain roof pitch.
- Drill .5-inch drain holes in the corners of the bottom (and elsewhere in the floor for Barn Owl boxes).
- Gouge or cut horizontal grooves into the inside of the front panel under the entrance hole to help young owls climb out.

The measurements in this table allow a .25-in. gap at the top for ventilation. For the Northern Saw-whet and Western Screech-Owl designs, the front panel is mounted with two pins that penetrate the side panels halfway between floor and roof, allowing the panel to swivel outward so the chamber can be cleaned. The panel is secured at the bottom with a screw.

siding works well. A box should be attached directly to a limb or tree trunk, with branches nearby that provide perches for the young owls as they leave the box and exercise their wings before they can fly. A few horizontal grooves routed or chiseled into the wood to about three inches under the entrance hole provide footholds for the birds. A strip of rubber or plastic can be used to protect the tree if the box is attached with wire rather than nails or screws.

Some box sizes suggested, particularly for Barn Owls, are too small. Although Barn Owls make use of nest boxes made for the similar-sized Wood Duck, these are really not large enough for a bird whose numerous young do not leave the nest until they are adult sized (baby Wood Ducks jump out of the boxes as tiny hatchlings, to be led by their mother to the safety of water). Eastern Screech-Owls were found to have slightly higher mortality and early fledging when they used the smallest of three sizes provided (Gehlbach 1994b).

Boxes may be mounted in trees or on sturdy posts. The "Bloom Box" is mounted on a horizontal tree limb by thick wires or cables. Creative box builders have tacked on bluebird boxes to the ends of rectangular Barn Owl boxes, thereby creating "duplexes." These "room additions" were promptly used by a variety of hole-nesting birds, such as Ash-throated Flycatchers *(Myiarchus cinerascens)*, which successfully raised broods despite the nocturnal grunts, shrieks, and screeches of the Barn Owl family next door (S. Simmons, pers. comm. 2007).

The Northern Pygmy-Owl generally does not make use of nest boxes in the United States, presumably because of the abundance of woodpecker holes (one such hole used by the owl measured 3.8 cm [1.5 in.] vertically by 5 cm [2 in.] horizontally). However, its Old World counterpart, the Eurasian Pygmy-Owl *(Glaucidium passerinum)*, bred in captivity, was successfully reintroduced in the Black Forest of Germany by providing boxes; German forests are generally so manicured that dead snags with cavities are not tolerated.

By installing artificial burrows in suitable habitat, Burrowing Owls have been successfully translocated from areas undergoing development. The owls are first confined in aviaries (hack pens) that enclose the holes and a patch of surrounding habitat and fed there for a period in order to acclimate them. However, juvenile Burrowing Owls moved by this method in Minnesota in an attempt at reintroducing the species into the western part of that state failed; not one of the 105 released owls eventually returned to breed (Martell et al. 2001). Two of five pairs similarly transplanted during the courtship period in California did, however, breed and produce young (Terrill and Delevoryas 1993). The better alternative, when possible, is to move the owls only short distances (less than 100 m [300 ft]) from their threatened burrows using a method called passive relocation. The installation of one-way doors prevents the owls from reentering their burrows, which encourages them to move on their own to artificial burrows built nearby in safe areas with adequate foraging habitat (Trulio 1995). In Idaho, moving nests with advanced nestlings short distances (72 to 258 m [236 to 846 ft]) away from construction sites to artificial burrows had limited success; some adults and young promptly returned to the natural (but threatened) burrow site (Smith and Belthoff 2001b), demonstrating the dogged persistence with which Burrowing Owls adhere to their nest burrow (Zarn 1974). However, passive relocation was used with great success at one central California airport, where owls, prevented from returning to old burrows near the runways, moved into both artificial burrows and natural squirrel holes at the edges of low-speed taxiways. This well-managed owl population increased dramatically, probably in part because of minimal exposure to people, dogs, cats, raccoons, and foxes. There was no concurrent rise in plane strikes (J. Barclay, pers. comm. 2007).

Captive breeding of owls may work but unfortunately does not guarantee that the offspring can successfully be released to the wild. Breeding programs of Barn Owls in Britain managed to produce 2,000 to 3,000 birds per year from the early 1970s to 1992, but the programs resulted in high postrelease mortality, with few providing birds that lived to breed (Bruce 1999).

Wildlife Rehabilitation

All through the western states, dedicated people working at rehabilitation centers take in injured or orphaned wildlife, in many cases donating their time as volunteers. Owls wind up at such centers with surprising frequency, most commonly as victims of vehicle collisions or as nestlings or fledglings. If the latter, conscientious rehabbers first ascertain that the birds are not merely normal prefledged young from an extant nest where they might

Fig. 90. Almost all owls normally prefledge, most before they can fly properly or at all. These Great Horned Owl fledglings perch together, as young owls frequently do. They should always be left alone unless they are in imminent danger.

be returned (human scent is of no consequence to birds). If they are indeed orphans, they are raised (which is normally not difficult) and allowed to catch live mice in a large flight cage in the dark and practice their foraging skills until they are judged capable hunters and released.

Injured or disabled owls are a different matter. Barn Owls especially may arrive severely malnourished and may have to be force-fed initially; if they eventually recover to feed themselves and to be released, their survival rate appears to depend upon the length of time spent in captivity, which influences loss of muscle mass and strength. Not surprisingly, those briefly confined do generally well, depending on how old and experienced they were before falling into human hands (M. Manhal, pers. comm. 2006).

It is very unlikely that the great majority of rehabilitations have any impact on the survival of a given species, with the possible exceptions of the California Condor *(Gymnogyps californianus)*, which would not be breeding or even surviving in the wild today without human intervention, and the Peregrine Fal-

Fig. 91. A wildlife rehabilitation center during an annual event releases rescued and rehabilitated owls.

con *(Falco peregrinus)*, which, after the use of DDT was banned, would have recovered on its own, although perhaps more slowly. However, the educational value of rehab centers is incalculable. There can be no doubt that they are a unique and important venue for acquainting the public with the local wildlife and thereby promote conservation at a time when commercial interests threaten to obliterate much of the West's wild heritage. Moreover, they provide developers with a means to salve their conscience (if present) by making donations to such enterprises.

SPECIES ACCOUNTS

Overview

Both common and scientific names of birds, as well as the taxonomic sequence used in this book, are in accord with the American Ornithologists' Union *Check-list of North American Birds,* seventh edition (1998), and, conforming to the University of California Press policy, the official common names are capitalized. Note that "Barn Owl" refers to the widespread species of owl found not only in North America but also on several other continents, but "barn owl" means any member of the barn owl family (there are 16). Similarly, although there are various species of pygmy-owls, only the Northern Pygmy-Owl *(Glaucidium gnoma)* is found in California and most of the West; but the Northern Spotted Owl *(Strix occidentalis caurina)* is merely a subspecies of the Spotted Owl. Occasionally, common names have been abbreviated, such as Short-ear, for Short-eared Owl *(Asio flammeus),* retaining the capitalized first letter to indicate reference to that particular species.

Each owl species is illustrated in color. Only the most commonly encountered form is shown. Two plates illustrate most of the young, at the late fledgling stage (which is when you would most likely come across them), of the species that nest in the western states and looks sufficiently different from the adults to pose an identification problem. Both adult and young are additionally described in the species accounts, with a quick identification synopsis provided for each species. In all species except the Burrowing Owl *(Athene cunicularia),* the female is larger than the male. Further, the sexes are usually alike or very nearly so in color and pattern; where the sexes are substantially different, both male and female are shown and described.

The most important criteria for the identification of owls are habitat, size, presence or absence of ear tufts, the color of the eyes, voice, and, lastly, plumage color. A small, leggy owl perched on the ground in grassland is certainly a Burrowing Owl, its identity confirmed by its tuftless head and yellow eyes. An even smaller, tuftless owl with yellow eyes, seen in a forest, with rusty streaks on its undersides, must be a Northern Saw-whet Owl *(Aegolius acadicus).*

Plumage pattern is very helpful in separating some species, as are tail length and size. It is assumed that the reader is sufficiently familiar with common birds—Red-tailed Hawk *(Buteo jamaicensis)*, Rock Pigeon *(Columba livia)*, American Crow *(Corvus brachyrhynchos)*, American Robin *(Turdus migratorius)*, European Starling *(Sturnus vulgaris)*—to make useful size comparisons.

The most often heard song or call of each species is described and transliterated where possible, as well as various other vocalizations. You can listen to calls online (for example, at www.owling.com or www.owlpages.com), or purchase audio guides from Cornell Laboratory of Ornithology (www.birds.cornell.edu).

Daily activity patterns and aspects of flight are provided in the species accounts to aid in identification, and sections on food habits and reproduction give insights into the lives of these often elusive birds. Some facets of owl biology, such as many aspects of behavior related to reproduction, hold true for all species of owls, and information concerning these can be found in the sections "An Owl's Body" and "An Owl's Life." Readers interested in learning still greater detail about owls' bodies and biology are encouraged to avail themselves of a surprisingly large body of scientific literature, some of it posted online, along with a number of fine books included in the reference list in the back of this volume.

Identification Key

Whereas length from tip of beak to tip of tail is the standard size measurement for birds, in the following key, the phrase "body size about that of a pigeon" means that the owl has roughly the *bulk* of a small pigeon, without being necessarily as long or as heavy, because much of the pigeon's length is supplied by its tail, neck, and beak, all of which, excepting the tail in some, are usually short in owls. The same applies to comparisons with robins, crows, and in Red-tailed Hawks. Bars are horizontal, streaks vertical.

Key to Adult Owls in the West

1a Eyes yellow or orange 5
1b Eyes dark, appearing black 2

2a Body size about that of a robin . . . . Flammulated Owl
2b Body size about that of a crow or larger. 3

3a Facial disk heart-shaped, whitish. . . . . Barn Owl
3b Facial disk like cut surface of halved apple, mottled 4

4a Breast and belly spotted . . . . Spotted Owl
4b Breast barred; belly broadly streaked, not spotted . . . . Barred Owl

5a Body size about that of a Red-tailed Hawk or larger. 6
5b Body size smaller than that of a Red-tailed Hawk. 8

6a Ear tufts conspicuous; body brown, mottled . . . . Great Horned Owl
6b Ear tufts absent 7

7a Body entirely white or white with black bars . . . . Snowy Owl
7b Body grayish brown; facial disk enormous with concentric rings . . . . Great Gray Owl

8a Body size about that of a crow 9
8b Body size smaller than that of a crow. 11

9a Ear tufts absent; belly finely barred; northern states . . . . Northern Hawk Owl
9b Ear tufts present; belly checkered or streaked; widespread 10

10a Ear tufts conspicuous; belly checkered. . . . Long-eared Owl
10b Ear tufts inconspicuous; belly streaked . . . Short-eared Owl

11a Body size about that of a small pigeon 12
11b Body size smaller than that of a pigeon 16

12a Ear tufts absent 13
12b Ear tufts present 14

13a Coniferous forests; legs short; facial disk large . . . . Boreal Owl
13b Open areas, terrestial; legs long; facial disk small . . . . Burrowing Owl

14a Song rapid, evenly spaced notes; Arizona, New Mexico only . . . . Whiskered Screech-Owl
14b Song a whinny or a bouncing ping-pong ball 15

15a Song a descending wail plus whinny; see range map . . . . Eastern Screech-Owl
15b Song like a bouncing ping-pong ball; see range map . . . . Western Screech-Owl

16a Body size about that of a robin; facial disk large . . . . Northern Saw-whet Owl
16b Body size much smaller than that of a robin 17

17a Belly with diffuse streaks; tail short; southern states only . . . . *Elf Owl*
17b Belly with clean streaks; tail long 18

18a Crown spotted; woodland and forests; widespread . . . . Northern Pygmy-Owl
18b Crown streaked; Sonoran Desert (Arizona only) . . . . Ferruginous Pygmy-Owl

Range Maps

Defining the exact ranges of owls is difficult. Although many owl species in the West are found throughout vast stretches of suitable habitat, others are more limited in their distribution and occur in specific habitats that are widely scattered across the landscape. Flammulated Owls *(Otus flammeolus)*, for instance, are frequently (though not always) associated with ponderosa pines *(Pinus ponderosa)* and are consequently, like the tree, found in disjunct populations across the western states. In addition, they tend to nest in clusters separated by habitat that looks identical to that used by the birds (Marti 1997b). During dispersal and migration, owls turn up in unexpected places. Flammulated Owls have been found on boats at sea, and the skeleton of one was discovered washed up on a beach near San Diego (P. Bloom, pers. comm. 2007), likely a migrant that lost its way while heading south. Some species are so hard to find that their ranges are poorly known. Great Gray Owls *(Strix nebulosa)*, for example, are difficult because not only are they the most secretive of all and disappear instantly, but also they often do not call until late at night, and they may not respond to taped calls in areas of low density; to find one, in some places this means hiking in a freezing forest at 2,400 m (8,000 ft) at one or two o'clock in the morning.

California distribution is emphasized; however, some species are so rare or occasional in California that defining a range is very difficult (although often easy in parts of the West where they are more numerous). Named counties are in California.

Range maps have been included for all species, but those for the Snowy Owl *(Bubo scandiacus)* (which does not generally nest south of the Arctic Circle) and the Northern Hawk Owl *(Surnia ulula)* (which only very rarely nests south of the Canadian border) are limited to delineating their southernmost western occurrences in winter.

Habitats are described in fairly broad strokes only, because owls seem to readily take to ecologic formations that are only superficially similar. Northwestern redwood forests and the mixed hardwood-conifer forests of Mount Shasta, for example, appear quite different to the human eye but are equally acceptable to species such as the Northern Pygmy-Owl and the Northern Saw-whet Owl. Owls are more specific about local amenities, such as plentiful voles, the presence of water, flowing or standing, and, for some, large numbers of woodpecker cavities suitable for nesting or, sometimes, roosting; others demand the presence of old stick nests built by other large birds.

BARN OWL — *Tyto alba*

Pls. 1, 20

FIGS.: 2, 7, 10, 18, 24, 28, 38, 40, 46–48, 54, 55, 56–58, 64, 69, 77, 87, 91–95

IDENTIFICATION: A slender mid-sized owl, about as large as a crow, with a heart-shaped facial disk. Its folded wings extend beyond the short tail, and its sparsely feathered, conspicuously long legs often look "knock-kneed," the result of perching with heels close together. The female is usually a bit larger than the male.

The unique facial disk, different in shape from that of all other North American owls, is white to buff, at times with indistinct reddish tan blotches, especially in the female. It encircles, like a skipper's beard, the long ivory bill. The eyes are small and dark.

The dorsal color of both sexes is ocher yellow to golden rust, with large patches of vermiculated gray and white feathers usually decorated with dark-rimmed white spots and dark brown spots. The underparts range from plain to heavily speckled, but those of the male are white to pale buff, whereas the female's are typically darker, from dark buff to yellowish tan or cinnamon.

There is no distinct juvenal plumage.

VOICE: The most commonly heard vocalization of the Barn Owl, given at night and in flight, is a hoarse, rasping "churrick," also rendered as "sshnaairkk," "shrreee," or "karr-r-r-r-ick." It has been described as resembling the sound of "a rake dragged across concrete" (D. Tiessen, pers. comm. 2006) and is chiefly given by males. This species, however, produces a wide variety of clicks, screeches, whistles, purrs, hisses, and chirrups, along with soft chattering and bill snapping. The male screams to announce his return with food to the nest.

FLIGHT: The moderately broad wings are long and usually very sparsely marked below, and the feet extend just past the short tail. The large, blunt head and thick neck project far forward. A Barn Owl typically glides and cruises with measured wingbeats at a height of about 3 m (10 ft) when foraging, although it may also fly at much higher altitudes, as attested to by the high mortality of this species in wind farms.

DAILY ACTIVITY PATTERN AND FEEDING: Chiefly nocturnal, the Barn Owl only occasionally hunts in the daytime. Much foraging takes the form of speculative flights over open country, including marshes, fields, and meadows, following established routes. However, it also still-hunts from perches such as telephone poles and light standards in cities, and it may chase prey on foot.

Although this species is a wide-spectrum feeder, catching animals from crickets—including Jerusalem Crickets *(Stenopelmatus fuscus)*—to midsized mammals, a cursory examination of a pellet pile under most Barn Owl roosts quickly confirms that in California and elsewhere, rodents make up by far the largest part of the diet and that, often, this raptor has a special fondness for pocket gophers *(Thomomys)*. It also takes many voles *(Microtus)* and mice and is strong enough to tackle ground squirrels *(Spermophilus)* and even jackrabbits *(Lepus)*. Overall, mammals comprise 73 to 100 percent of this owl's diet. Birds are also taken (though usually in small numbers), normally small and medium-sized songbirds such as European Starlings *(Sturnus vulgaris)*, but also waterbirds up to the size of American Coots *(Fulica americana)*. One pair of Barn Owls raised its young on an apparently steady diet of Leach's Storm-Petrels *(Oceanodroma leucorhoa)* on an island off California's coast (Bonnot 1928). In addition, lizards, snakes, frogs, and large arthropods (such as scorpions in desert areas) may appear on the menu. The Barn Owl, being a pragmatist, catches what is available within a certain size range.

REPRODUCTION: The male Barn Owl may live in his breeding territory year-round but defends it against other males only during the breeding season, this species being only mildly territorial (if at all, some think). If prey is plentiful, defended areas can be as small as a few meters around the nest, thus allowing for semicolonial breeding.

Barn Owls begin to breed typically as one-year-olds and often do not survive long enough to breed again. The male initiates the process by calling from a prospective nest site to attract a mate and by engaging in advertising flights with stiff, deliberate wing-beats and wing-clapping. Ardent suitors, males accumulate numbers of dead rodents to impress a prospective bride. A willing female is pursued in the air with wails that recall those heard during nocturnal cat adventures. The male then entices the female with calls and flights to enter the nest cavity.

The nest site is very frequently a human-made structure such as a barn, silo, bridge, steeple, mineshaft, culvert, or hole in a building. Clearly human-friendly, Barn Owls readily utilize nest boxes where such are provided for them. Natural nest sites include the "skirts" of palms, cracks and pockets in cliffs, caves, and badger holes. Holes in the vertical dirt banks of arroyos, at times dug by the owls themselves (Millsap and Woodruff 1979), are sites commonly used in the foothills flanking the Central Valley of Cal-

Fig. 92. Barn Owls took advantage of a break in the roof edge of an old water tower to find a nest site. European Starlings saw the opportunity, too, and used the same hole for nesting.

Fig. 93. Hollows and pockets in sandstone cliffs, particularly with an old Common Raven's *(Corvus corax)* nest in place, are popular with Barn Owls.

ifornia. Tree holes, too, make fine homes, especially those of oaks *(Quercus)* in savannas and oak woodlands, probably more often than is generally realized. Beyond a slight depression, no nest is built, although the female breaks up her own pellets to form a soft substrate for the oval or near-spherical eggs (W. Winsted, pers. comm. 2005).

Barn Owls produce many young and, reproductively, quickly respond to high prey availability, but mortality is also high. They lay their eggs from February (in southern California [Hanna 1954]) until June, but one observer found eggs in every month of the year (P. Bloom, pers. comm. 2006). Depending on the abundance of food, a female Barn Owl may lay up to 15 eggs, although four to seven are most common. She may produce two clutches (or more, especially on other continents) if prey is plentiful. In southern California, for example, about 10 percent of the pairs have three broods during peak rodent years. As many as eight pairs nested in a barn at Camp Pendleton (San Diego County), with eggs or young of all ages (P. Bloom, pers. comm. 2006).

Incubation lasts about a month. Eggs are laid at two- to three-day intervals, and the female begins to sit after laying the first or

Fig. 94. Barn Owls are more widespread in the margins of oak woodland and in riparian woodland than is generally realized.

Fig. 95. A Raven's nest in a rarely used barn holds young Barn Owls close to fledging.

the second egg. As much as three weeks may separate the hatching of the first and last young, and a typical nest has chicks in various stages of development and in widely different sizes. Fledging is therefore protracted but occurs at about nine weeks of age. When food is scarce, the smaller chicks may become food for their larger siblings, which may kill them, and the adults may eat their own small young when gophers and mice run out, although probably after they have starved to death.

Larger young break up pellets and kick them, together with fecal material, out of the nest cavity if possible (Marti 1992).

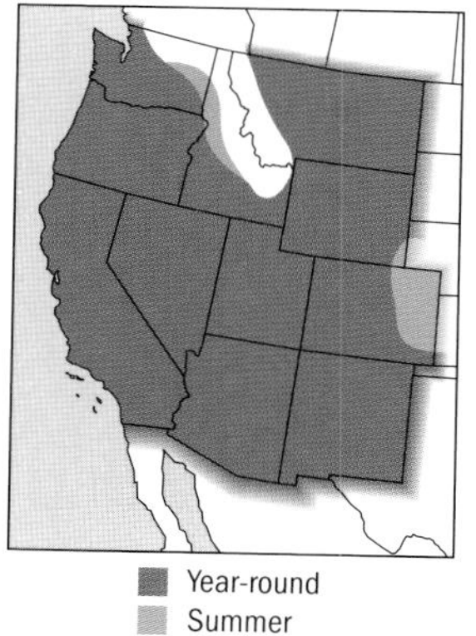

DISTRIBUTION AND HABITAT: Nearly cosmopolitan, the Barn Owl occurs on all continents except Antarctica, and of the world's land birds, it is one of the most widespread; as many as 46 races have been described, with the North American the largest in size. The species is found nearly throughout the Southern Hemisphere, but in the Northern Hemisphere it is confined to the temperate to tropical zones; it is largely absent from Canada and scarce in the northern parts of its range. Whereas the Barn Owl does not do well in cold climates, populations in some temperate and tropical regions are expanding their ranges because of increases of suitable habitats and use of nest boxes.

The Barn Owl is probably the most often encountered owl in California and most other western states, frequently as a roadside casualty. It occurs almost throughout California, generally at elevations below 4,500 ft, including most of the Channel Islands, avoiding high mountains and dense forests. It is found in a variety of habitats but requires open ground for foraging. Typical habitat includes roost sites adjacent to or included in grasslands, open fields, beaches, marshes, deserts, and even lawns and gardens. Nest sites may also serve as roost sites, as does the dense foliage of willows *(Salix)*, oaks, pines *(Pinus)*, planted redwoods *(Sequoia)*, and eucalypts *(Eucalyptus)*. Barn Owls are frequently heard in towns and cities at night and may loiter around prey-attracting street lights.

California's Barn Owls do not appear to move very great distances. Individuals banded in Modesto (Stanislaus County) were recovered across the Sierras in Bridgeport and Bishop (Mono and Yolo counties), and another flew to Santa Cruz (Santa Cruz County) (S. Simmons, pers. comm. 2005). Although this species appears as a migrant on Cape May, New Jersey, and occasionally may move far, it is usually sedentary where it occurs in North America (unless dispersing after the nesting season), with few individuals moving generally southward in winter.

SIMILAR SPECIES: Most frequently seen at night, the Barn Owl is commonly (and incorrectly) perceived as very large and completely white, and so it is often mistaken for a Snowy Owl *(Bubo scandiacus)*, an exceedingly rare species in California that is much larger and more massive and either nearly all white or white with blackish bars both above and below. The Short-eared Owl *(Asio flammeus)* is even longer winged than the Barn Owl and has conspicuous underwing markings; it is more active in the daytime and is not as pale as the Barn Owl.

STATUS: Although the Barn Owl is on the endangered species lists of some midwestern states and is a Montana state species of concern, it has no special conservation status in any other western state. Numbers have, in fact, increased in California and British Columbia because of increased nest site availability and range expansion (DeSante and George 1994). Populations are less secure in Wyoming, Idaho, and Utah.

REMARKS: This species is also known as the Common Barn Owl, Barn-Owl, Monkey-faced Owl, and White Owl.

A cornered Barn Owl puts on an impressive threat display. Not only does it, like other owls, bend forward and spread its wings to increase its apparent size, but it also may then go on to rock from side to side and back and forth and swing its head like a pendulum, or lunge at the intruder, all the while hissing like escaping steam, snapping its bill, and stamping its feet.

On a warm spring night in the inner Coast Range, two of these owls swooped repeatedly at two young dogs scampering by a campfire, occasionally hurling invective in the form of blood-curdling shrieks.

FLAMMULATED OWL *Otus flammeolus*

Pls. 2, 21

FIGS.: 25, 96

IDENTIFICATION: A very small owl (the only small owl with dark eyes), about robin sized, brownish gray, variably marked with rufous, and with blackish shaft streaks and crossbars. The colors are matched to local tree bark. Its small ear tufts are often laid flat. The tail is short, but the wings are long.

The Flammulated Owl has two forms: rufous (red) and gray, with many intermediates. The head is gray brown, with varying amounts of rufous, and with paler gray eyebrows. The facial disk is gray, paler toward its outer border, where it is framed dark brown or blackish and rufous. Additional rufous appears around, and particularly above, the eyes, which have bumpy, brown to reddish lids that give the eyes a rheumy appearance.

The owl's upperparts are dark gray and brown or rufous, with brown black shaft streaks and fine markings. The outermost scapulars are adorned with bold white spots all in a row, bordered with broad ocher to rufous margins. The breast is gray, blending into white on the belly and flanks. Fine brown black transverse bars cover the underparts, along with bold blackish shaft streaks that widen toward the flanks and are variably bordered with rufous. The tarsi are densely feathered, and the toes are naked. The feet look oddly small and delicate for an owl.

The juvenile is grayish white barred with brown black. The facial disk has broad dark borders and lightens toward the pink-lidded eyes, which, like the adult's, look rheumy. The wing and tail feathers look like those of adult owls.

VOICE: The advertising song of the male Flammulated Owl is a distinctive "boop-boop-boodle boop," best imitated in a low falsetto. The pitch is approximately B above middle C (Marshall 1939). The call is repeated over and over, the singer falling silent only when he moves to another perch, typically high and close to the trunk of a big tree. Males respond to imitated calls from after their arrival in April or later until mid-June (i.e., during the mating period) all night long as well as at dawn and dusk. Unmated males may sing throughout the summer.

A hoarse, rushing sound—like that of a nighthawk *(Chordeiles)* in its nuptial dive—is produced by a male when another

Fig. 96. Long wings benefit long-distance migrants such as Flammulated Owls.

from a neighboring territory alights near it. The female utters a long, single "oooo," softer than the male's "boop" (Marshall 1939). Additional Flammulated Owl vocalizations include shrieks, screams, moans, whining, and gurgling sounds. The young "snore" and peep.

FLIGHT: The Flammulated Owl is long winged and flies strongly, useful traits for a long-distance migrant. It flies swiftly between trees, makes abrupt turns, and resembles a bat in long zigzag flights (Marshall 1939). Point-to-point flight is straight, not undulating.

DAILY ACTIVITY PATTERN AND FEEDING: Strictly a nocturnal hunter, the Flammulated Owl spends the day roosting on branches next to the tree's trunk so that its patterned plumage melds with the bark. It also crams itself into bark or tree clefts but does not roost in cavities such as woodpecker holes (except for breeding females). At all times, this owl strives to conceal itself by choosing perches in dense vegetation, and even a singing male seeks out song perches in the highest and densest trees available, stationing himself close to the trunks, surrounded by foliage, close to but never at the tree's top. Even while hunting, this owl habitually alights within the trees (Marshall 1939). Foraging is done in the form of flycatcherlike sorties from a perch, resulting in skillful capture of flying prey. This owl also hovers while gleaning prey from foliage or picks it up from the ground.

Arthropods comprise by far the major part of the Flammulated Owl's diet, with lepidopterans (mostly moths), crickets and grasshoppers (orthopterans), and beetles (coleopterans) making up the bulk. The long debate whether this species ever kills vertebrates has been settled by observations at nest boxes in northern Utah, adding to previous evidence; here, mice, bats, and songbirds or their remains were found. In one box, the observers discovered an owl youngster with a deer mouse *(Peromyscus)* halfway down its gullet. However, of 1,875 prey deliveries videotaped over 93.3 hours, only one was of a deer mouse, brought by the male to the brooding female (Oleyar et al. 2003).

REPRODUCTION: Male Flammulated Owls arrive on their breeding grounds earlier than females and, having established territories, seek to attract the latter upon their arrival by persistent singing. The tardy females, on their part, move through the males' territories uttering food-begging calls (Reynolds and Linkhart 1987). A successful male takes his mate (which may be the same as the previous year's if she has survived) on a tour of potential nest cavities; presumably it is she who selects one. Typical sites include cavities, such as those made by Northern Flickers *(Colaptes auratus)* and Pileated Woodpeckers *(Dryocopus pileatus),* and nest boxes, where made available. The owls are very fond of holes in dead ponderosa pine *(Pinus ponderosa)* snags. In a New Mexico study, most adults did not use a tree cavity again even though they had successfully nested there the previous year (McCallum et al. 1995).

Being migrants, Flammulated Owls begin nesting relatively late, as owls go, and egg-laying in California and other western states occurs in May and June and perhaps as late as early July. A relatively long-lived species, Flammulateds are like large raptors in that they have a relatively low annual reproductive rate (Linkhart and Reynolds 2006). A typical clutch consists of just two or three relatively big eggs (four-egg clutches are rare), which are laid on whatever debris the cavity holds; no nest is built. Incubation lasts three to three and a half weeks and is carried out by the female, which is only slightly larger than the male, this species showing the least sexual dimorphism of all North American owls (Earhart and Johnson 1970).

The young are fed exclusively with food the male delivers. At the age of 12 days, they are left alone as the female joins the male in foraging, and about three to three and a half weeks after hatching, the owlets fledge. As in other owls, fledglings may leave the nest well before they can fly. The parents provide food for another month or so after fledging, at which time the young begin to disperse.

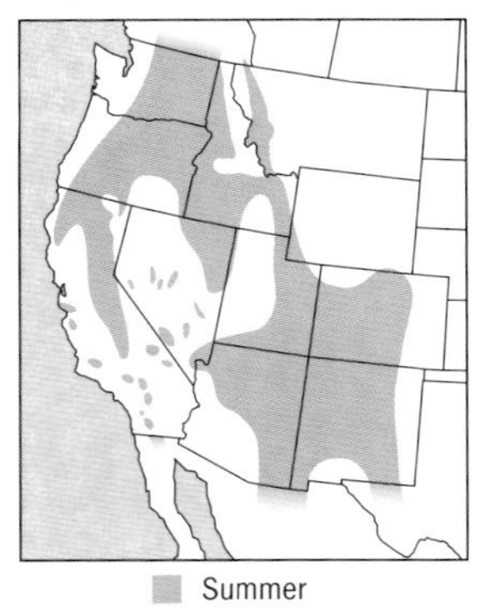

DISTRIBUTION AND HABITAT: The Flammulated Owl breeds from southern and southeastern British Columbia south through nearly all of Washington and Oregon, very limited areas of Montana and Idaho, and from there in an irregular band through the western states into Mexico down to the Isthmus of Tehuantepec and the northern Baja Peninsula.

Whereas Flammulated Owls may turn up in some unexpected places on migration in California (they are occasional vagrants to Santa Barbara Island, for example), they occur as breeders in the northwest and in isolated populations in some of California's higher Coast Ranges (as far south as Monterey County, in black oak–ponderosa pine habitat, above 1,150 m [3,800 ft] with most above 1,500 m [5,000 ft]), the Transverse Ranges, and Peninsular Ranges and southern mountains. The main population, however, is found across the Cascades (where it is most common in mixed conifers) and down the length of the Sierra Nevada. Here, the birds occur commonly at elevations from 370 to 1,700 m (1,200 to 5,500 ft), but also up to 3,050 m (10,000 ft). Nesting occurs at higher elevations in the southern part of this owl's range than in the northern part.

Breeding takes place usually in rather open forests that have high arthropod populations, have some dense foliage for roosting, and are cold and often dry (McCallum 1994b). Although this species is associated with coniferous forests in general, it is especially partial to the higher parts of open ponderosa pine or Jeffrey pine *(Pinus jeffreyi)* forests, often mixed with black oak

(Quercus kelloggii) and with a shrubby understory. In a Colorado study, breeding pairs clearly preferred ponderosa pine–Douglas-fir *(Pseudotsuga menziesii)* forests that were 200 to 400 years old (Linkhart and Reynolds 1997). The partiality to ponderosa pine and Douglas-fir habitats may be based on the abundance of lepidopterans found in these trees (McCallum 1994b), four times as many as in other common western conifers (Furniss and Carolin 1977).

At higher elevations, it is found in red fir *(Abies magnifica)* forests. Aspens *(Populus tremuloides)* and mixed forests of white fir *(Abies concolor)*, Douglas-fir, incense-cedar *(Calocedrus decurrens)*, and sugar pine *(Pinus lambertiana)* also are suitable habitat if open in aspect and containing shrubs and dense vegetation. Nest boxes placed in aspens (among some scattered firs) were used for breeding in northern Utah, with the migrant Flammulateds sometimes finding the boxes already occupied upon their return by Northern Saw-whets *(Aegolius acadicus)* (Marti 1997b). In southwestern Idaho, however, these birds arrived earlier than did migrating Northern Saw-whets and were in better physical condition, as befits long-distance migrants (Stock et al. 2006).

Flammulateds are the most migratory of our owls, and next to nothing is known of their lives in their winter quarters south of the border other than that they seek out the same habitat as at home. Trapped on migration in Arizona and New Mexico, they appear to pick different migratory routes in fall and spring because of prey availability (Balda et al. 1975).

SIMILAR SPECIES: The Flammulated Owl has diminutive ear tufts and is much smaller than the relatively long-tufted screech-owls *(Megascops)*, which it otherwise resembles. Above all, its eyes are dark, not yellow like those of the screech-owls. There is little overlap in habitat of the Flammulated with these. Likewise, the very small Northern Saw-whet Owl also has yellow eyes, in addition to a distinctive, at times angular head (without tufts) and broad, rust brown ventral striping. The Northern Pygmy-Owl *(Glaucidium gnoma)*, similar in size to a Flammulated, is yellow eyed, too, and has a noticeably long tail.

STATUS: In appropriate habitat, this species is common. Unfortunately, ponderosas and other conifers are of great interest to the logging industry, leading the U.S. Department of Agriculture to designate the Flammulated Owl a sensitive species, the bird having shown a decline in numbers after timber harvesting. Also, cutting down snags and diseased trees for firewood removes suitable nest sites. Wildfire suppression, too, has a negative effect on this owl's habitat. Pesticides sprayed on forests may decrease the arthropod food available to the owl, and European Starlings *(Sturnus vulgaris)* compete for the same nest sites, along with American Kestrels *(Falco sparverius)*, Western Screech-Owls *(Megascops kennicottii)*, and Northern Saw-whet Owls, all of which may beat the migratory Flammulated to available nest cavities.

The breeding population is considered vulnerable in Idaho. In Montana, it is a state species of concern, and it is listed as a stewardship bird species in Nevada.

REMARKS: Compared to other small owls, the Flammulated appears rather delicate. It is readily preyed on by some diurnal raptors, such as the Northern Goshawk *(Accipiter gentilis)* and the Cooper's Hawk *(A. cooperii)*, as well as larger owls.

Handled owls may become catatonic and remain motionless when placed on the ground (Reynolds and Linkhart 1998), and incubating females may be sluggish and unresponsive (Richmond et al. 1980).

This owl used to be called the Flammulated Screech-Owl.

WESTERN SCREECH-OWL *Megascops kennicottii*

Pls. 3, 21

FIGS.: 8, 17, 19, 25, 30, 50, 66, 97–99

IDENTIFICATION: A small owl about the size of a small pigeon with a very short tail, with usually obvious ear tufts on its big head that may, however, be laid flat. The eyes are conspicuously yellow.

In most of its range, the head and upperparts (including upperwings) are gray to gray brown, finely vermiculated with paler gray and with variably narrow dark brown streaks. Brown and even rufous brown birds occur in the coastal Pacific Northwest, with the reddest individuals found in coastal Washington and Oregon; the palest light gray birds are found in the deserts of the lower Colorado River.

The facial disk is pale gray, usually with three or four indistinct concentric darker gray rings around each eye, and it is bordered by black brown crescents on either side. The eyebrows formed by

Fig. 97. An example of the pale race of the Western Screech-Owl *(Megascops kennicottii yumaensis),* lured one morning from its hole in a desert ironwood *(Olneya tesota)* by the author's whistle.

the disk are conspicuously paler than the rest and often whitish, the white usually extending to the inside margins of the tufts. A white triangular bib with fine dark marks below the beak is flanked by blackish bars. The color of the beak is dark gray or blackish, with a pale yellow tip.

A row of large white spots extends down the margin of the scapulars, with a second, disrupted row on the outer upper secondary coverts. The breast and belly are whitish, with black brown streaks of variable width and variably fine crossbars, especially on the flanks. The legs are thickly feathered, brownish gray, sometimes with darker bars, and the toes are sparsely feathered.

The juvenile is loose feathered in appearance, with wispy tufts appearing above the eyes. The head and underparts are densely patterned with narrow gray brown bars on a paler gray to buff background. The back and upperwings are intricately patterned in gray, ocher, and black brown, with a few white accents. Overall, a young Western Screech-Owl looks like it is wearing an old-fashioned striped jail suit.

More or less distinctive local forms of this owl are recognized as nine separate subspecies (four of which, plus intergrades, occur in California and six in the western states) that vary in weight and color or both. The relationships between screech-owls as a group are not fully understood, and DNA analysis may reveal additional full species.

VOICE: The most often noticed vocalization is the "bouncing ball" call, similar to the accelerating cadence of a ping-pong ball dropped on a hard surface. This is the advertising song given by both sexes (with the male having a lower voice). Double trills, with the first shorter than the second, are sung by both sexes. A uniform string of hoots, reminiscent of those of a Northern Saw-whet Owl *(Aegolius acadicus)*, is produced by males during courtship. Additional vocalizations include peevish-sounding alarm barks as well as drawn-out whines, chuckles, and occasional screeches. Older nestlings, fledglings, and females beg with short whinnylike descending trills.

FLIGHT: The wings appear broad and very rounded when the owl flies overhead, and the flight is direct and fairly rapid.

DAILY ACTIVITY PATTERN AND FEEDING: The Western Screech-Owl normally spends the daylight hours roosting in tree cavities or nest boxes, thickets, and vine tangles such as poison-oak *(Toxicodendron diversilobum)*, next to tree trunks and elsewhere in trees, and in abandoned cabins and in caves and crevices. One owl, occupying a nest box, would regularly hang halfway out of the hole as the box was heated daily by the early afternoon sun in summer. Although this behavior was clearly an effort to cool off, screech-owls may perch in the entrance of their roost cavity during sunny winter days to warm up (Cannings and Angell 2001); even when there is no sun, some owls sometimes lodge themselves in their nest box entrances seemingly keeping a half-open eye on their surroundings.

This owl leaves its roost to forage shortly before or shortly after sunset (Hayward and Garton 1988). It hunts almost entirely from perches but is capable of hawking flying insects and capturing bats in flight.

The Western Screech-Owl is nearly as catholic in its prey choices as the very closely related Eastern Screech-Owl *(M. asio)*. In addition to the above, the Western Screech-Owl also hunts birds, captured at their night roosts, some as large as Northern Flickers *(Colaptes auratus)* and Steller's Jays *(Cyanocitta stelleri)*, and even a female Mallard *(Anas platyrhynchos)* (Cannings and Angell 2001) has been recorded, which, however, very probably was injured. Much more commonly, this owl eats arthropods such as Jerusalem Crickets *(Stenopelmatus fuscus)*, ants, and crayfish (caught in shallow water), as well as earthworms. Small fish, salamanders, and reptiles can form part of the diet, too. In some areas, rodents predominate on the menu, including fairly large ones such as wood rats *(Neotoma)* and pocket gophers *(Thomomys);* the remains of three adult cottontails *(Sylvilagus)* were found in a nest box (Cannings and Angell 2001). Obviously, the Western Screech-Owl helps itself to whatever is locally and seasonally available.

REPRODUCTION: Although a Western Screech-Owl defends its territory throughout the year, advertising songs become louder and longer during the breeding season (Herting and Belthoff 1997). The size of territories varies with prey abundance and nest site availability. In Orange County, 14 territories were found along a riparian corridor of 6.4 km (4 mi) (Feusier 1989). Polyterritorial, this owl defends alternative nest holes (Gehlbach and Gehlbach 2000).

Fig. 98. This Western Screech-Owl has moved into a Wood Duck *(Aix sponsa)* box holding duck eggs and added her own. (Photo by Steve Simmons.)

These owls are monogamous and pair members frequently engage in allopreening, which reinforces the pair-bond. The male presents his mate with food and uses the latter to lure the female into nest cavities. Copulation is preceded by allopreening, perching together shoulder to shoulder, and by antiphonal singing and calling. Typical nest sites are woodpecker *(Colaptes* and *Dryocopus)* holes, natural tree cavities, and nest boxes. Occasionally, the owls may use pockets in the "skirts" of dead fronds of palms (Cannings and Angell 2001).

Nest chambers are typically from about 3 to nearly 9 m (10 to

Fig. 99. As the young grow, nest boxes often get very crowded. (Photo by Steve Simmons.)

nearly 30 ft) above ground level. The white oval eggs, laid from the end of February to early April, usually number three to five, although clutches of nine have been found. Incubation is carried out by the female and lasts from a little less than four weeks to a bit more. The male roosts a few dozen feet from the nest in dense cover and feeds the female at night throughout incubation and beyond. She joins him in foraging once the young are about three weeks old. Soon the chicks scramble up to the nest hole and, well before they leave the cavity, inspect their home grounds. They fledge at night, four to five weeks after hatching and, having very poor powers of flight, remain near the nest site for several days.

DISTRIBUTION AND HABITAT: The Western Screech-Owl is found from Alaska south through western British Columbia, the Pacific states into Baja California through the Great Basin, and from the western Rocky Mountains states south through Colorado and western Texas into western Mexico and the Mexican interior highlands.

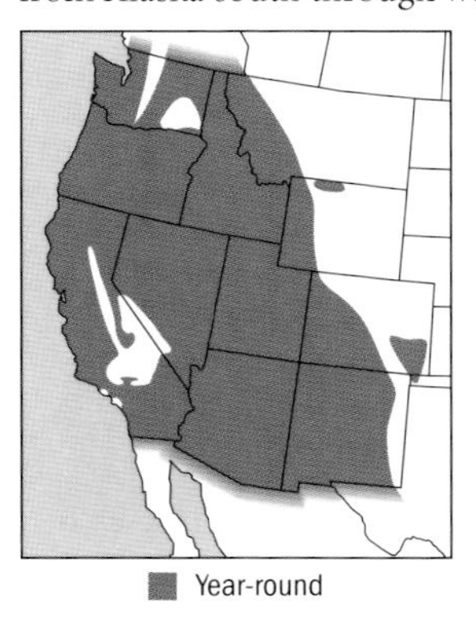

Widespread in California and elsewhere, it can be found from near sea level to above 2,400 m (8,000 ft) (Garrett and Dunn 1981). It is, however, absent from the crest of the Sierra Nevada, the Warner Mountains, the Cascade-Sierra axis, the White Mountains, the central and western Mojave, the Salton Basin, as well as parts of coastal southwestern California (Small 1994) because of habitat loss.

Breeding occurs in a variety of habitats, ranging from riparian woodland (including desert washes) and open oak woodland (where it is perhaps most common) to piñon-juniper woodland, to redwood and mixed hardwood and conifer forests (Zeiner et al. 1990). This owl does not fear people and is found in suburbs and even older city parks (it needs woodpecker holes in mature trees), as well as around farms.

Western Screech-Owls are nonmigratory, and pairs remain on their breeding grounds year-round.

SIMILAR SPECIES: The two owls most similar to the Western Screech-Owl—the Eastern Screech-Owl and the Whiskered Screech-Owl *(M. trichopsis)*—do not occur in California, but the latter can be sympatric with the Western in Arizona. The three species are so similar that without the voice, they are usually impossible to tell apart, although the Whiskered is clearly smaller than the other two. Besides the bird's song, beak color is said to be the only field mark that distinguishes this owl from the Eastern Screech-Owl. However, beak color is highly variable, particularly in Easterns, and likely not very visible anyway. Most helpful is knowing the ranges of the three species, which, in addition to the voice, may verify a given owl's identity.

One other owl in the West that resembles the Western Screech-Owl a bit is the Flammulated Owl *(Otus flammeolus)*, which is not only much smaller but also has tiny (instead of large) ear tufts and, most importantly, dark (not yellow) eyes. The remotely similar Long-eared Owl *(Asio otus)* is nearly twice as large and has a rusty or ocher facial disk marked with black and white. The Northern Saw-whet Owl is smaller than the Western Screech-Owl, and, along with the similar-sized Boreal Owl *(Aegolius funereus)*, although yellow eyed, it lacks ear tufts and has a distinctively flat, wide facial disk.

STATUS: As elsewhere, California's Western Screech-Owls are probably slowly declining in numbers, chiefly because of habitat destruction. Particularly damaging is the elimination of thousands of acres of open oak woodland to make room for urban development and agricultural development (including viticulture) and for conversion into firewood. Additional hazards include vehicle collisions, the invasion of the Barred Owl *(Strix varia)* (which, upon breeding in the area, caused the departure of Western Screech-Owls near Vancouver, British Columbia [Cannings and Angell 2001]), and trichomoniasis, a protozoan infection that is frequently fatal and that the owls most likely contract from diseased prey. Still, the Western Screech-Owl is probably the most numerous owl in the wooded West, with its mellow bouncing calls adding charm to a warm spring night.

REMARKS: A captive female, inadvertently imprinted on humans, had the run of the house and, although very tame and affectionate with the host family, fiercely attacked human visitors, vexing them with sometimes wounding blows to the head when their backs were turned and forcing them to accept proffered hats and helmets, made available to them at the door. The owl, upon seeing her handler, would often pop into her nest box (actually, a log section with a natural cavity hung high on the wall), and scratch about in the bottom in a seductive fashion. At times she would back into her hole, modulating her call, which is considered to be male behavior. Eventually, she went on to lay a few infertile (of course) eggs. Like some other captive Western Screech-Owls, this bird lived for 19 years, longer than those in the wild.

EASTERN SCREECH-OWL *Megascops asio*

Pls. 4, 21

FIG.: 100

IDENTIFICATION: A small owl, about the size of a small stubby-tailed pigeon, with usually conspicuous ear tufts (when erect) and yellow eyes. The beak is yellowish or greenish (see the Western Screech-Owl *[M. kennicottii]* species account, "Similar Species" section).

The Eastern Screech-Owl occurs in two morphs, the gray and the rufous, and there are many intermediates between the forms.

The head and upperparts of the gray form are gray to gray brown, with fine vermiculations, especially on the head, and with variably wide blackish shaft streaks crossed by curving narrower bars, forming anchor mark shapes. The facial disk is grayish and has concentric dark or gray rings, and the prominent white of the eyebrows continues up to the inner margins of the ear tufts. The disk is framed on both sides by blackish brackets that continue to the sides of the throat, which is whitish and finely barred and streaked. A row of white spots marks the outer border of the scapulars, with a second disrupted row on the outer secondary coverts. The breast and remainder of the underparts are white to grayish white, with variably wide shaft streaks forming anchor marks. The legs are whitish or buff.

The rufous morph, which comprises 36 percent of the population throughout its range (15 percent in the West), is essentially bright rusty rufous where the other form is gray or gray brown. The facial disk is entirely rufous or rufous and white, the eyebrows and rictal bristles white to off-white, and the head and back rufous with simple blackish shaft streaks lacking crossbars. There are large white spots on the outer scapulars and a second disrupted row on the upper secondary coverts. The underparts are bright white with blotches (some oak-leaf shaped) or broad crossbars of rufous bordering blackish shaft streaks of variable width. The legs are rufous or white.

The juvenile Eastern Screech-Owl is loose feathered and fluffy, grayish to buff, with rudimentary ear tufts and intricately vermiculated black brown and buffy to ocher upperparts. The facial disk is grayish, with darker concentric circles, and the underparts are finely and profusely barred, the bars most dense on the breast.

Five subspecies of the Eastern Screech-Owl are recognized, defined by color and size.

VOICE: The advertising song of the Eastern Screech is a descending wail followed by a whinny, a series of downward-inflected quavering notes or single note. A monotonic trill is used as a pair or family contact song; tremulous and bouncy, the song may rise or fall. Other vocalizations are hoots, indicating low-level alarm concerning intruders and predators, barks, which indicate a higher alarm level, and screeches, given during defense of the young. Additional calls include rasps and a chuckle-rattle (Gehlbach 1995).

FLIGHT: As in other small owls, the flight of the Eastern Screech-Owl is direct, curving only to avoid obstacles. When flying from one habitual perch to another, it drops off the first, flies low to the ground, and abruptly swings up to the second. It can also briefly hover. Rarely gliding, this owl flies with rapid wingbeats.

DAILY ACTIVITY PATTERNS AND FEEDING: During the nesting season, male Eastern Screech-Owls roost in dense deciduous foliage or vine tangles within 6 m (20 ft) of incubating females, closer after

Fig. 100. Like some other owls, a red-phase Eastern Screech-Owl "broods" a frozen mouse to thaw it for consumption.

the young hatch. In winter, they use cavities, nest boxes, or conifers, selecting south- and west-facing sites (Gehlbach 1995).

This owl sets out to forage shortly after sunset, earlier in cloudy weather. A still-hunter, it quickly moves from one perch to the next. It drops on or flies down to prey, even into water, and flails its quarry from foliage with wingbeats. This owl also catches insects in the air with its beak (Sutton 1929).

Of all the small owls, the Eastern Screech-Owl preys on the widest range of animals. Invertebrates such as earthworms, snails, crayfish, beetles, crickets, and moths are commonly taken, and, from the ranks of vertebrates, a variety of fish, frogs, toads, and salamanders, a great assortment of lizards and snakes, along with a broad selection of birds (including other Eastern Screech-Owls) and mammals, chiefly rodents but also small rabbits and bats. Vile taste or smell, as that of ground beetles, and spikey armor do not deter the owl (Gehlbach 1995).

REPRODUCTION: Like some other small owls, the male defends several nest cavities and the surrounding areas (polyterritories). The female selects one of these, encouraged by the male's gift of dead mice and by previous nesting success. Natural cavities in trees such as apple trees *(Malus)*, oaks *(Quercus)*, pines *(Pinus)*, and elms *(Ulmus)* include hollow trunks and limbs, old woodpecker holes, and rotted-out stumps. Among human-made sites are nest boxes, mailboxes, buildings, and boxes on the ground, but normally higher.

The two to seven (typically three or four) eggs are incubated by the female only for a little over four weeks, the young prefledging after another four weeks or so, earlier in larger, more crowded broods (Gehlbach 1994a). As in other owls, the male provides food for the female all through incubation and hatching until she no longer broods the young. Flightless for the first few days after leaving the nest, the chicks are attended by the parents for another month or more.

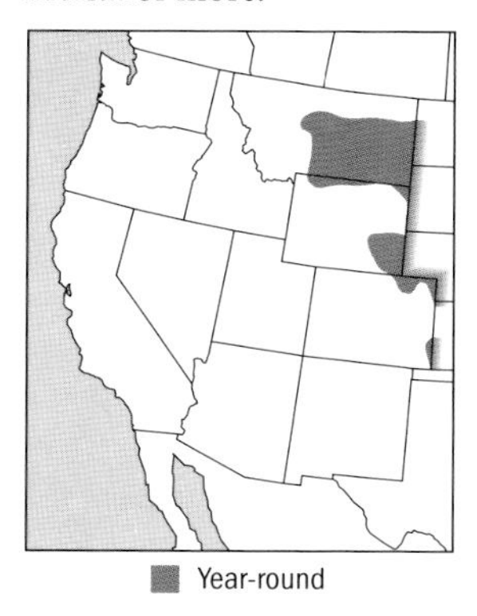

DISTRIBUTION AND HABITAT: Almost entirely a bird of the eastern half of the United States, the western border of this owl's range is in the Great Plains; in the western states, it is found only in parts of Montana, Wyoming, and Colorado.

The Eastern Screech-Owl inhabits areas with trees chiefly below 1,500 m (4,900 ft). Natural or human-altered woodlands or forests of deciduous hardwood or pine are equally as acceptable as riparian woodlands, city parks, playgrounds, greenbelts, and backyards.

Sedentary, the species does not migrate, though juveniles disperse, as do adults when food is scarce.

SIMILAR SPECIES: The Western Screech-Owl is extremely similar to the Eastern. In the field, only the voice is a near-certain identifier—there is some slight overlap in the vocalizations. Fortunately, the two species are sympatric in only very few places in west Texas and perhaps northern Wyoming. The Whiskered Screech-Owl *(M. trichopsis)*, though similar, does not occur in the Eastern's range (except in Mexico), and the Flammulated *(Otus flammeolus)* is much smaller and has dark eyes, its range barely overlapping with that of the Eastern in west Texas.

STATUS: Although sensitive to habitat loss through tree removal, Eastern Screech-Owls show only cyclic population fluctuations, the result of inconstant food availability. By and large, the species is highly adaptable to human landscapes and can often be encouraged by the placement of nest boxes in reforested areas and suburbs.

REMARKS: The Eastern Screech-Owl can be fierce in its nest defense and does not hesitate to strike humans, sometimes drawing blood, when it considers its young threatened.

Rufous individuals freely pair up with gray ones, and males can be polygynous, maintaining more than one mate.

WHISKERED SCREECH-OWL *Megascops trichopsis*

Pls. 5, 21

FIG.: 101

IDENTIFICATION: A small owl, smaller than a pigeon and other North American screech-owls, with usually prominent ear tufts (which, however, can be flattened on the large head) and conspicuously golden yellow (or orange) eyes.

The whiskerlike disk feathers indicated by this owl's name are virtually impossible to see in the field. The gray facial disk has a black brown border that is in turn edged with rusty buff, and the eyebrows formed by the disk's inner margins are conspicuously pale or white, as are the inner edges of the ear tufts. A buffy or rusty patch covers the throat.

Where it is found in the United States, this species is generally gray (but there is a rufous morph in Mexico, its center of distribution, and southward). Darker gray vermiculations and dark brown to black shaft streaks and fine crossbars (anchor marks) decorate the body feathers both above and below. A row of irregular white spots runs down the outer scapular margins, with similar but shorter lines along the edge of the folded wing and over the upperwing coverts. The upper legs are rust colored, the lower white.

The juvenile Whiskered Screech-Owl is narrowly but faintly crossbarred gray on head and body, and its feathers have the same loose texture seen in other young owls.

Of the three races, only one is found in the United States. The identification of this owl is greatly facilitated by its very limited range in the United States, and its distinctive and unique song.

VOICE: One advertising song of the male, announcing the presence of a territory and of a nest cavity, has been given a variety of names by a half dozen authors who appear unable to agree on a common term. It is a series of notes, "ook-ook-ook . . . ," with the last dropping in pitch, delivered fast but evenly spaced, without speeding up like the song of the sometimes sympatric Western Screech-Owl *(M. kennicottii)*. Variations of this song are used for emphasis, inter- and intraspecific defense, male-female interactions, and other purposes (Gehlbach and Gehlbach 2000). Calls (defined as notes separated by intervals more than twice as long as the notes and, unlike the songs, that lack harmonics) include hoots, whistles, barks, screeches, squeaks, squawks, and more (Gehlbach and Gehlbach 2000).

FLIGHT: This owl is a skillful flyer that turns and darts through a woodland canopy and below. Its foraging flights, starting from perches, are swift and direct.

DAILY ACTIVITY PATTERN AND FEEDING: The Whiskered Screech-Owl roosts during the day well out of sight, with nesting males usually electing to perch close to tree trunks or, about equally often, in dense foliage. It also roosts sometimes in cavities, locations chosen by nesting females. Both sexes sunbathe in cavity holes, particularly following rain or a cold night.

Foraging begins about dusk, the first bout lasting under an hour and, while there are nestlings to feed, goes on irregularly all through the night until dawn, when there is one last burst of effort. This owl often flutters back and forth in treetops catching insects, scarcely pausing to rest (Holt et al. 1999). Hunting flights are direct. Starting from a perch, the owl pursues prey in foliage, on the ground, or in the air. Whereas the quarry is usually caught with the feet, aerial captures may be accomplished with the beak.

Unlike other screech-owls, the Whiskered's diet focuses, though not exclusively, on arthropods (particularly while nesting, when these comprise the main prey); many moths are taken, as well as caterpillars (even hairy ones), and beetles can feature prominently. Vertebrates from lizards and snakes to shrews, bats, and mice are also prey, the larger animals sometimes stored (cached) whole or in halves, particularly when there are nestlings.

REPRODUCTION: Syncopated duetting of a pair of Whiskered Screech-Owls led to copulation, which was followed by bill rubbing and allopreening (Smith et al. 1982). The female selects the cavity from among those shown her by the male, typically an old woodpecker hole in an Arizona sycamore *(Platanus wrightii)* (though other trees may be used), about 8 m (25 ft) up, sometimes lower, in Arizona riparian woodland at an elevation of 1,500 to 1,750 m (4,900 to 5,700 ft). She lays most often three (but from two to four) eggs, which she alone incubates, probably for 26 days, the male providing her with food. As in other owls, he continues to feed the family after the young hatch and are brooded by the female until she resumes hunting. The young fledge prematurely over a period of about a week, apparently induced, at least in part, by the female's whistling, and require from three to four days to learn how to fly short distances (Gehlbach and Gehlbach 2000).

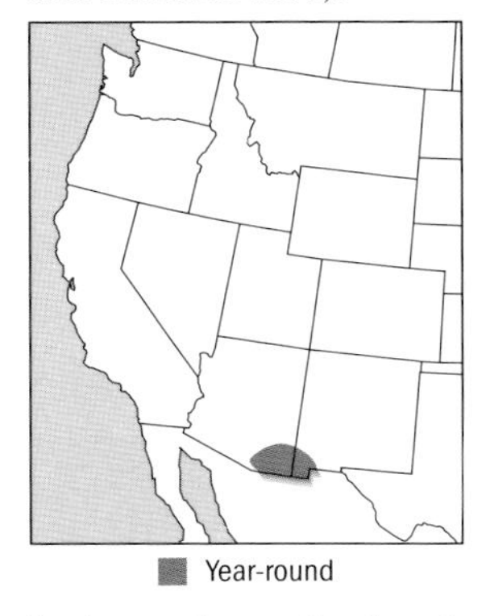

DISTRIBUTION AND HABITAT: In the United States, this owl is found only in the mountains of southeastern Arizona and extreme southwestern New Mexico, whence its range extends southward through the mountains of interior Mexico to the south of that country, with a disjunct population occurring in El Salvador, Honduras, and northern Nicaragua. The Whiskered Screech-Owl inhabits evergreen pine-oak woodland or mixed juniper-oak woodland at elevations of 1,000 to 2,900 m (3,300 to 9,500 ft), with Arizona sycamore and Arizona walnut *(Juglans major)* in riparian habitat preferred for breeding. Unafraid of people, the Whiskered Screech-Owl nests sometimes in campgrounds.

They are sedentary although they may move to lower elevations in winter.

SIMILAR SPECIES: Overall, this owl is very similar in appearance to the gray forms of both the Eastern Screech-Owl *(M. asio)* and the Western Screech-Owl, although significantly smaller (by about an inch). Although the Eastern's range in the United States does not overlap that of the Whiskered, that of the Western does, and the size difference may not always be readily apparent in the field; although the Whiskered is found usually at higher elevations, the two species can be told apart with certainty in the field only by their voices. The Whiskered has eyes of a deeper yellow than a Western, but this difference is not always obvious. The less similar Flammulated Owl *(Otus flammeolus)* is not only much smaller but also has dark eyes; it can be found, however, in the same habitat as the Whiskered.

STATUS: Apart from disturbances at the nest by birdwatchers, Whiskered Screech-Owls seem to hold their own, although food shortages and severe weather can impact the U.S. population. Replacements, however, are probably supplied by the Mexican population. Small numbers of this screech-owl were first found in Hidalgo County, New Mexico, in 1990, and worries about loss of habitat from fire (including prescribed burns) led that state to list it as threatened.

REMARKS: Upon imitating an unfamiliar owl song in southeastern Arizona in March many years ago, I was pleasantly startled to have a Whiskered Screech-Owl land on the ground six feet in front of me, apparently quite prepared to see off the presumed trespasser. Other observers have reported similar encounters. However, such responsiveness works to the owl's detriment. Because of its very limited distribution in the United States, birders from throughout the country converge on this owl's habitat to add the species to their life lists. In one Arizona canyon, the number of birders between 1987 and 1992 tripled so that an estimated 2,100 visitors per month inflicted themselves upon the owl during April and May, the peak nesting months, resulting in fewer or less fit young (Gehlbach and Gehlbach 2000).

In hand, the Whiskered Screech-Owl has noticeably small feet, probably reflecting its insectivorous food habits.

Fig. 101. In response to the playing of a taped song, male Whiskered Screech-Owls may land on the ground before an owler to challenge the perceived intruder.

GREAT HORNED OWL *Bubo virginianus*

Pls. 6, 20

FIGS.: 14, 15, 21, 23, 24, 27, 29, 38, 41, 57, 60, 81, 82, 88, 90, 102, 104

IDENTIFICATION: A very large owl that appears about the size of a Red-Tailed Hawk but is actually often heavier (females average about 1.6 kg [3.5 lb] while the average female Redtail weighs a bit over 1.1 kg [2.5 lb]). Long black ear tufts bordered with ocher on the inside are usually conspicuous and, when seen in frontal silhouette, make the head appear a bit like that of a cat. The huge eyes are pale to deep orange yellow. The female is larger than the male.

The gray and brown head is finely vermiculated to coarsely blotched with black brown, a color and pattern resembling bark, which continues over the entire upperparts. The facial disk is grayish, pale ocher to reddish brown and bracketed with black crescents; the rictal bristles are whitish. The upper eyelids are conspicuously white, as is the throat. The eyebrows are pale ocher.

The upper and central breast is white, blending into ocher toward the sides and belly, with black blotches of the sides of the upper breast and with narrow transverse bars on the lower breast, belly, flanks, and underwing and undertail coverts. Legs and toes are thickly covered with pale ocher or buff feathers nearly to the talons. The fleshy soles are naked and studded with tiny scalelike tubercles.

The juvenile is often ragged in appearance, with short ear tufts. The contour feathers, except for the flight feathers, are downlike, whitish buff to grayish in color, with narrow darker grayish brown transverse bars and often with down from earlier coats adhering to their tips. The remiges are well grown by the age of about three weeks, but the tail lags behind. The facial disk is similar to that of an adult, as are the eyes.

Five subspecies occur in the western states, all of them apparently also recorded in California, differing in degrees of color saturation, size, and more subtle criteria. The darkest birds are found in the northwest, the palest in the deserts. The largest subspecies, and killing and go through the usual owl contortions to assess distances. By the age of about six weeks, and well before they can fly, the youngsters move out of the nest onto nearby branches and ledges (Houston et al. 1998). They can fly at about seven weeks, though not yet very well. The fledged young tend to stay together and beg from (and are fed sporadically by) the adults well into fall, although the latter typically roost elsewhere, probably to escape the incessant begging. They finally disperse in fall and, if no breeding territories are available, become nonnesting floaters, sometimes for several years.

Most Great Horned Owls, especially the females, vigorously defend their nests against perceived predators, including humans, and can inflict painful injuries and shred clothing.

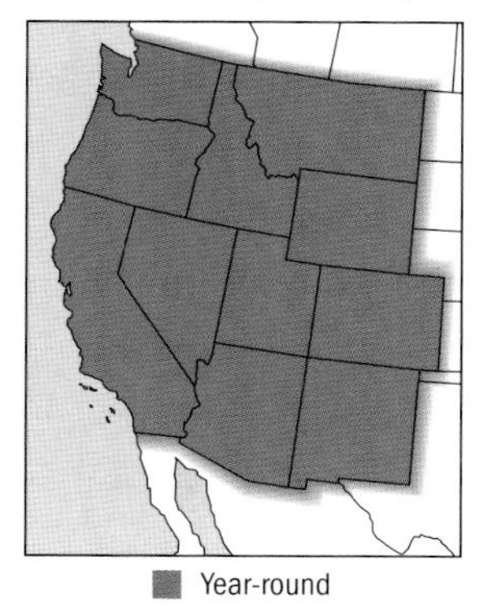

Year-round

DISTRIBUTION AND HABITAT: The range of the Great Horned Owl extends from northern Alaska and central Canada across all of North and Central America, to northern South America and central and eastern parts of that continent.

In California and elsewhere in the West, it is the most widespread (though not the most numerous) of all our owls. With the exception of the highest parts of the Sierra Nevada crest, it is found the length and breadth of the state in a great variety of habitats, sometimes even making it to Santa Barbara Island. The species is perhaps most commonly encountered in oak savanna, open oak woodland, lower-elevation coniferous forests with open areas for foraging, and most rarely in dense high-elevation coniferous forests. It can be expected, however, in orchards as well as in chaparral, riparian woodland, and parks and suburbs with large, dense trees, and it is not uncommon in deserts. Thickly wooded canyons provide good habitat for roosting and nesting, and barns in farmland are also acceptable. As many as 61 individuals were recorded in one night on the Monterey Peninsula (Roberson 2002). In the sage country of the Great Basin, they seek out dense stands of piñon pine *(Pinus)* or juniper *(Juniperus)* or large rock outcrops.

SIMILAR SPECIES: Only two other owls are the size of a Great Horned Owl, and both lack ear tufts. The Snowy Owl *(B. scandiacus)* is conspicuously all white (or nearly so), and the Great Gray Owl *(Strix nebulosa)* (which, because of its long tail, appears larger than the Great Horned) has a distinctive enormous facial disk and, like the Snowy, lacks ear tufts.

The considerably smaller Long-eared Owl *(Asio otus)* lacks the barred belly and has relatively much smaller eyes, and its ear tufts arise much closer to the head's midline than do those of the Great Horned Owl. Compared to the latter's bulk, it appears slender and delicate, and its head looks a bit pointed, unlike the flat, wide tiger head of the larger cousin.

STATUS: California and the rest of the West have a healthy population of Great Horned Owls, as reflected, sadly, by the relative frequency with which this species shows up as a road casualty. It is adaptable to habitat changes as long as these do not interfere with an adequate prey base, and provide nest and roost sites.

REMARKS: The Great Horned Owl, as a broad-spectrum predator, utilizes a greater range of habitats than any other western owl, and because it does not object to the proximity of humans, the nests at times become local attractions.

The resonant calls—chiefly heard in late fall, winter, and spring—are familiar to many people. The call is beloved of movie makers who wish to convey just how utterly lost the protagonists are in the dark woods, though, oddly, the gentle call of the Eurasian Little Owl *(Athene noctua)* seems to be preferred for night scenes shot near dwellings, even for movies set in North America, where the species is absent.

This species is the proverbial Hoot Owl. It is also known as the Cat Owl.

PLATES

Relative Sizes of Western Owls

a. Great Gray Owl (68.5 cm [27 in.]), plate 15

b. Snowy Owl (58.5 cm [23 in.]), plate 7

c. Great Horned Owl (56 cm [22 in.]), plate 6

d. Northern Hawk Owl (40.5 cm [16 in.]), plate 8

e. Long-eared Owl (38. cm [15 in.]), plate16

f. Short-eared Owl (38 cm [15 in.]), plate 17

g. Barred Owl (53.5 cm [21 in.]), plate 14

h. Spotted Owl (44.5 cm [17.5 in.]), plate 13

i. Barn Owl (40.5 cm [16 in.]), plate 1

j. Boreal Owl (25.5 cm [10 in.]), plate 18

k. Burrowing Owl (24 cm [9.5 in.]), plate 12

l. Western Screech-Owl (21.5 cm [8.5 in.]), plate 3

m. Eastern Screech-Owl (21.5 cm [8.5 in.]), plate 4

n. Whiskered Screech-Owl (18.5 cm [7.25 in.]), plate 5

o. Feral Rock Pigeon (32 cm [12.5 in.])

p. Northern Saw-whet Owl (20.5 cm [8 in.]), plate 19

q. Northern Pygmy-Owl (17 cm [6.75 in.]), plate 9

r. Ferruginous Pygmy-Owl (17 cm [6.75 in.]), plate 10

s. Flammulated Owl (17 cm [6.75 in.]), plate 2

t. Elf Owl (14.5 cm [5.75 in.]), plate 11

u. European Starling (21.5 cm [8.5 in.])

a
b
c
d
e
f
g
h
i
j
k
l
m
n
o
p
q
r
s
t
u

Barn Owl, PAGE 162

Medium size (40.5 cm [16 in.]), long legs, dark eyes, heart-shaped disk, and generally pale coloration. Underwings virtually unmarked. Darkest individuals are typically females.

Widespread, and at times and in places common. Often associated with human-made structures. Chiefly nocturnal.

Male

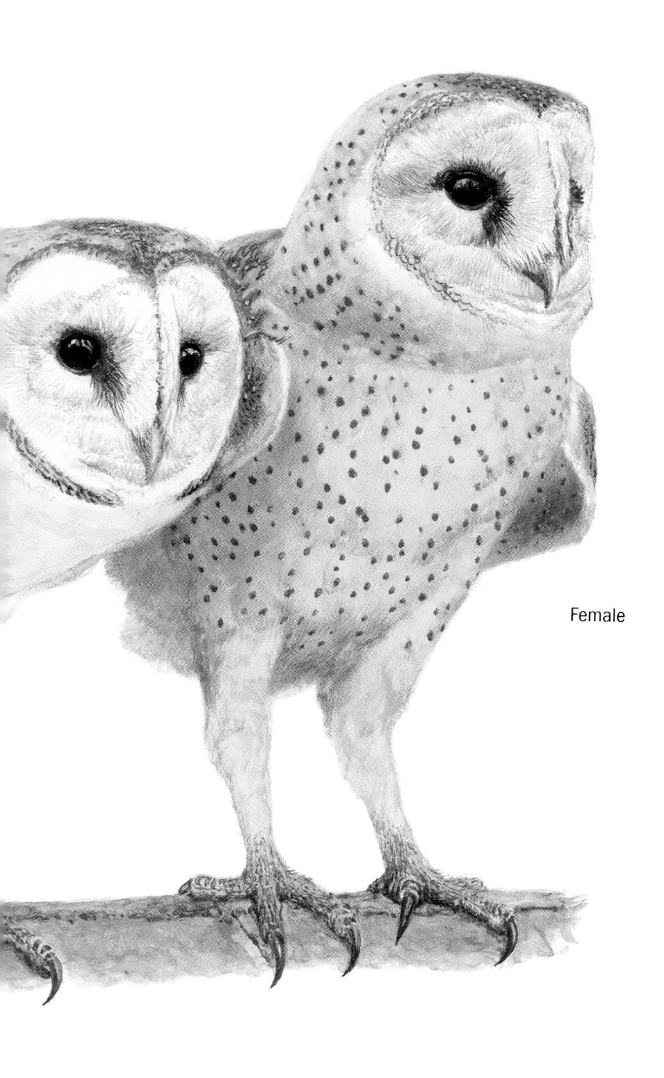
Female

Flammulated Owl, PAGE 169

Small size (17 cm [6.75 in.]), dark eyes, and small ear tufts that are often hard to see; well-developed facial disk. Typically has varying amounts of rusty coloration.

Restricted to mountains, common in places, often associated with ponderosa pines.

PLATE 2

Western Screech-Owl, PAGE 174

Small size (21.5 cm [8.5 in.]), yellow eyes, and long ear tufts (shown depressed here). Underparts finely streaked and barred. Northwest coast individuals may be reddish brown.

Widespread and common; found in a variety of habitats including parks and suburbs.

PLATE 3

Eastern Screech-Owl, PAGE 180

Indistinguishable from the Western Screech-Owl in the field except by voice and except for its red form, which is unique to this species in the United States.

Widespread and common east of the Rocky Mountains; in the West found only in parts of Montana, Wyoming, and Colorado.

Red form
Gray form

Whiskered Screech-Owl, PAGE 184

Small size (18.5 cm [7.25 in.]) (noticeably smaller than Western Screech-Owl and Eastern Screech-Owl), ear tufts relatively smaller than in other screech-owls, deep yellow eyes. Colored and patterned like other screech-owls. "Whiskers" cannot be seen in the field.

Uncommon and found only in mountains of extreme southwestern New Mexico and extreme southeastern Arizona.

PLATE 5

Great Horned Owl, PAGE 188

Large size (60 cm [22 in.]), wide head, yellow eyes, and conspicuous, widely spaced ear tufts. Underparts uniquely (for big owls) finely cross-barred.

Widespread and common; found in a variety of habitats, from forests to deserts and suburbs.

PLATE 6

Snowy Owl, PAGE 195

Large size (58.5 cm [23 in.]), yellow eyes, predominantly white. Back of head of first-year male is mostly white, whereas first-year female has back of head marked with black; first-year females are much more heavily marked than first-year males.

Rare and irregular winter visitor in the coastal and northern states of the West, where it seeks out dunes, marshes, grasslands, and agricultural lands.

First-year male

Adult male

Northern Hawk Owl, PAGE 200

Medium size (40.5 cm [16 in.]) (mostly because of its long tail), with a flat, strikingly patterned head, yellow eyes. Tail appears wedge shaped in flight. Hawklike.

Rare and irregular winter visitor, very rarely a breeder in some northern states.

PLATE 8

Northern Pygmy-Owl, PAGE 204

Very small size (17 cm [6.75 in.]), yellow eyes, poorly developed facial disk. Head finely spotted. A brown horseshoe pattern appears on the upper breast and sides, distinct streaks on breast and belly. Long tail always barred black and white.

Widespread but uncommon in forested areas.

PLATE 9

Ferruginous Pygmy-Owl, PAGE 213

Size (17 cm [6.75 in.])and shape similar to Northern Pygmy-Owl, yellow eyes. Head finely streaked. Underparts finely or broadly streaked. Long tail may be barred black and white, rufous and dark brown, or plain rufous.

Rare in the West; found only in low desert of southern Arizona.

PLATE 10

Elf Owl, PAGE 217

Tiny size (14.5 cm [5.75 in.]), yellow eyes, absence of ear tufts, back marbled grayish, diffuse broad streaks on breast and belly. Short tail.

Uncommon but more numerous than generally recognized; normally found only in southern New Mexico and southern Arizona, possibly extreme southeastern California.

PLATE 11

Burrowing Owl, PAGE 221

Small size (24 cm [9.5 in.]), yellow eyes, small facial disk, long legs, and variably barred underparts. Generally loam colored. Female is more heavily barred than male and is smaller than the male.

Widespread but local in grasslands, agricultural areas, and industrial parks where these border open lands.

PLATE 12

Spotted Owl, PAGE 229

Medium size (44.5 cm [17.5 in.]), dark eyes, and large, well-developed facial disk. Spots cover the head and body (even the bars on the breast surround large white spots). Chocolate brown to tan, depending on race, the Northern Spotted Owl (shown) being the darkest, the Mexican Spotted Owl being the lightest.

Uncommon; found only in the coastal western states and in Colorado, New Mexico, Utah, and Arizona, in forests.

PLATE 13

Barred Owl, PAGE 235

Medium size (53.5 cm [21 in.]), dark eyes, well-developed, pale facial disk. Unique combination of barred upper breast with the rest of the underparts streaked.

Uncommon; found in parts of Montana, Idaho, Washington, Oregon, and California. Generally prefers wetter habitats than the Spotted Owl.

PLATE 14

Great Gray Owl, PAGE 240

Very large size (68.5 cm [27 in.]), small yellow eyes, elaborate facial disk with conspicuous bulls-eye pattern. Long tail. Overall grayish brown in color. Indistinct bars and streaks on underparts.

Rare to uncommon in Montana, Wyoming, Idaho, and the northwestern states; very rare in central and northern California.

PLATE 15

Long-eared Owl, PAGE 247

Medium size (38. cm [15 in.]), yellow to orange eyes, long and closely adjoined ear tufts, colorful facial disk, slender body, checkered underparts. Perches in trees, usually well concealed.

Widespread but uncommon throughout the West, in woodlands, forests, and riparian habitats near open country. Nocturnal.

PLATE 16

Short-eared Owl, PAGE 253

Medium size (38. cm [15 in.]), yellow eyes ringed with black or dark brown mask, ear tufts very small and rarely visible. Mottled brown and buff upperparts, streaked underparts. Often sits on ground, fenceposts.

Widespread but uncommon throughout the West, in grasslands, marshes, and other open lands.

PLATE 17

Boreal Owl, PAGE 261

Small size (25.5 cm [10 in.]), yellow eyes, well-developed facial disk bracketed with dark brown or black, heavily spotted wide head and more sparsely spotted upperparts, coarsely streaked underparts.

Uncommon or rare in subalpine forests of Montana, Wyoming, Colorado, northern New Mexico, Idaho, and very limited areas of Washington and Oregon. Unrecorded but suspected in California.

PLATE 18

Northern Saw-whet Owl, PAGE 265

Small size (20 5 cm [8 in.]), yellow eyes, conspicuous facial disk with sunburst pattern. Streaked head, some spotting on upperparts; underparts broadly streaked reddish brown. Short tail.

Widespread but uncommon to fairly common in closed-canopy woodlands and forests of the West.

PLATE 19

Fledglings of Large and Midsized Owls

Note distinctive facial disks. The Snowy Owl, which nests only in the Arctic, is not shown.

a. Long-eared Owl

b. Short-eared Owl

c. Northern Hawk Owl

d. Barn Owl

e. Spotted Owl

f. Barred Owl

g. Great Gray Owl

h. Great Horned Owl

PLATE 20

Fledglings of Small Owls

The fledglings of the Western Screech-Owl, Eastern Screech-Owl, and Whiskered Screech-Owl are alike in color and pattern; the last is noticeably smaller than the others. Not shown are fledglings of the Northern Pygmy-Owl and Ferruginous Pygmy-Owl, which are very similar to the adults.

a. Elf Owl

b. Flammulated Owl

c. Western, Eastern, or Whiskered Screech-Owl

d. Burrowing Owl

e. Boreal Owl

f. Northern Saw-whet Owl

PLATE 21

SNOWY OWL *Bubo scandiacus*

Pl. 7

FIGS.: 105, 106

IDENTIFICATION: A very large owl, without apparent ear tufts (though it has rudimentary ones), that appears like a stuffed pillow (owing to its density of feathers), almost all white or variably narrowly barred brown, tan, or blackish. The eyes are deep yellow and rimmed with black.

The adult male is either pure white or very sparsely, narrowly barred. The adult female shows moderate to extensive barring. A first-year female is more heavily barred, and the adult female, though chiefly white, has some barring on the breast, back, wings, and sometimes head or tail. There is considerable variation in degree of marking in females, less so in males. In both sexes, the dense rictal bristles nearly cover the black beak, and the legs and toes are densely feathered down to the black talons. It is the heaviest of all North American owls.

The juvenile Snowy is "dark mouse brown" (Parmelee 1992)

Fig. 105. Snowy Owls have the most pronounced rictal fans, which develop in the fledgling period (fledgling shown at left).

above and below, with some of the feathers tipped pale gray so as to give the bird a partly frosted appearance. The scapulars, upperwings, remiges, and tail are white with blackish bars, representing the first-year plumage, and the chin and facial disk are white, with black eyeliner below the eyes and black triangles above.

VOICE: Largely silent in its winter quarters (where it may on occasion scream in territorial defense), the Snowy Owl, particularly the male, becomes very vocal when breeding. The primary song is a double "hoo hoo" or a variably long series of "hoos," the last being the loudest. Far carrying, the booming hoots of these owls were heard from 11 km (7 mi) away across a harbor (Sutton 1932). Other vocalizations include an outburst of barking calls, given by both sexes but usually by the male, and calls that are described as rattling, mewing, cackling, and melodious warbling (Parmelee 1992). The prefledged young, hiding in vegetation, utter shrill squeals, presumably to aid adults in locating them (Parmelee 1992).

FLIGHT: The flight of this species is strong and steady, with a jerky upstroke of the wings and with intermittent gliding (Sprunt 1955), and is reminiscent of the flight of a buteo *(Buteo)* or harrier *(Circus)* because of the quick upstroke and slow downstroke (Watson 1957). At speed, while hunting birds in the air, the flight of this owl suggests that of an enormous falcon.

DAILY ACTIVITY PATTERN AND FEEDING: The Snowy Owl is mostly diurnal, on its wintering as well as on its breeding grounds (where daylight may be almost continuous at times). It forages chiefly by still-hunting, waiting on a perch until prey is located with its very keen eyesight. It may also hover and detect concealed prey not only by sight but also by acute hearing. A powerful flier, this species can also catch its quarry after direct chases in midair or take it off the water and on the ground (Parmelee 1992).

On its breeding ground, the Snowy Owl feeds on lemmings when available, virtually exclusively so during lemming peak years. An adult owl is estimated to consume three to five of these rodents daily (Watson 1957). This species also preys on hares and rabbits of various species, mustelids, and an assortment of rodents, as well as birds including songbirds, ptarmigan, geese, ducks and grebes (the latter snatched off the water), and sea birds. The Snowy Owl regularly eats carrion where available, such as the remains of slaughtered seals, dead Arctic Foxes *(Alopex lagopus)*, or a dead Walrus *(Odobenus rosmarus)* (Voous 1989).

REPRODUCTION: Snowy Owls may not breed every year, depending on the food supply. Males announce breeding territories they have established by hooting, at times in concert with neighboring males, and defend these territories aggressively. The owls are usually monogamous.

The male's courtship display includes a showy display flight during which he often carries a dead lemming in his beak or feet, presumably demonstrating not only his hunting prowess but also the presence of a food supply. This is followed by a ground display that may lead to copulation. Oddly, males may fly after a female disturbed from the nest and, upon landing, copulate with her, a type of displacement activity (Watson 1957).

A scrape in the ground on a hillock or hummock serves as a nest. The eggs, laid at two-day intervals, number from three to 11, with few eggs (three to five) the rule when food is scarce. The female develops a floppy, large, bare brood patch on her belly for incubation, which lasts about four and a half weeks. While lying on the nest, she is remarkably inconspicuous, especially with some remnant snow on the hillock but where it does not accumulate because of wind.

Females have been seen to feed freshly regurgitated pellets to the bigger young (Robinson and Becker 1986), suggesting a nutritious, perhaps predigested content. At the age of two to three weeks, the young leave the nest but are still far from flight. They are fed by the parents for at least an additional five weeks, when they start to forage on their own (Watson 1957), though they initially do poorly and require supplemental feeding. Nearly 7 kg (15.5 lb) of lemmings is consumed by a young owl in these five weeks, so that a brood of nine will have eaten as many as 1,500 lemmings by the time the parents stop feeding them (Parmelee 1992).

DISTRIBUTION AND HABITAT: The Snowy Owl is circumpolar. In North America, it breeds from the western Aleutian Islands in a variably narrow zone across western and northern Alaska and the Canadian Arctic including the Canadian Arctic Islands, east to northern Newfoundland.

The breeding habitat in North America consists of arctic tundra covered with lichen and vegetation a few inches high, including willows. Winter habitat in the northern Great Plains resembles the breeding habitat (Kerlinger et al. 1985).

On migration, this species regularly visits the central northern Great Plains in winter, although there is much local variation.

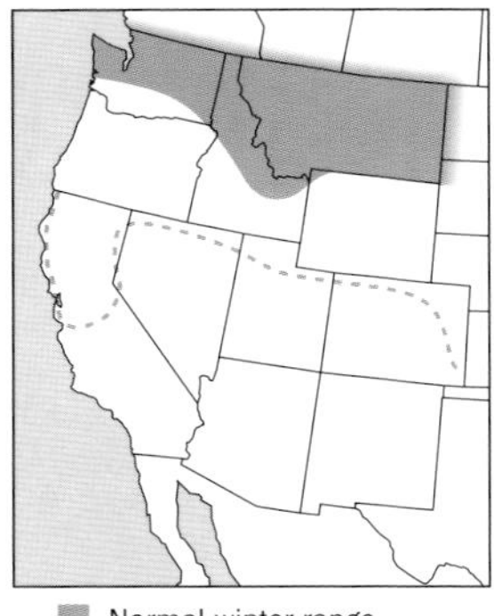

Normal winter range
Southernmost occurrences in irruption years

Overall, there is a three- to five-year pattern of appearances of Snowy Owls farther south in the continent (during so-called irruption or invasion years), synchronous with crashes in lemming population cycles, which, however, are likely not uniform throughout the range of the owl, making the bird's winter movements and distribution inconsistent (Kerlinger and Lein 1988). Immature male Snowy Owls generally venture farthest south, whereas adult females, having pushed the smaller males out, spend the winter in the northern plains regions. Owls invading states to the east and west of the Great Plains tend to be immatures (Kerlinger et al. 1985).

When the owl reaches the northwestern states and northern California, it normally seeks out coastal beaches, dunes, salt marshes, and river shores near the coast. Inland, it visits country resembling its tundra home. Large and conspicuous, it roosts in the open on migration, often on fence posts, road signs, and other elevated perches, less commonly in trees, but also on the ground, as the owl does at home on the tundra.

Snowy Owls very rarely visit California, even in irruption years. The last big winter invasion was in 1974, with some sightings including the San Francisco Bay Area and as far south as Monterey County (Small 1994). In January 2006, one Snowy appeared in the agricultural Delta area of Solano County, the first recorded sighting in California in 28 years, an event that drew hundreds of birders during the few days it lingered (B. Power, pers. comm. 2006), and another visited the same area in 2007 (G. and M. Trabert, pers. comm. 2007).

SIMILAR SPECIES: No other large white owl exists. The midsized Barn Owl *(Tyto alba)* may appear white but has in fact ocher brown upperparts, and it is almost entirely nocturnal and does not, unless forced, roost in the open like a Snowy Owl.

STATUS: Populations seem to be declining in Western Europe (Voous 1989), but elsewhere, this species is not threatened.

Fig. 106. On migration, Snowy Owls seek out terrain similar to the tundra, such as sand dunes and marshes.

REMARKS: Although they or their young may become prey for Snowy Owls, some birds, particularly Snow Geese *(Chen caerulescens)* often place their nests very near those of the owls (notwithstanding plentiful nesting habitat elsewhere). Apparently, the protection provided by the aggressive owls against ground predators such as foxes, make this gamble worthwhile. However, much like falconiform raptors, the Snowy Owl is said to rarely pursue animals close to its own nest (Parmelee 1972).

Native Americans in parts of Alaska may legally hunt these owls with no bag limit as long as they are used for food or their skins for clothing. Some consider the Snowy Owl to be the best-tasting raptor.

NORTHERN HAWK OWL *Surnia ulula*

Pls. 8, 20

FIGS.: 107, 108

IDENTIFICATION: A very distinctive midsized owl, about the size of a crow, with a long tail and a large, flat head. Most un-owl-like, it sits by day on prominent perches (such as treetops) in the open, commonly in a diagonal position, not upright like other owls. The eyes are small and bright yellow.

The brownish black head, back, and folded wings are heavily spotted; seen from behind, the bird looks perennially enveloped in snow flurries. The small, gray white facial disk is bordered by conspicuous black borders, wide on the sides, and there are a black bow tie under the beak and a black V-shaped pattern on the nape, reminiscent of the "false eyes" of pygmy-owls *(Glaucidium)*. The wings are short but pointed, and the brown black, graduated tail has seven to eight white crossbars.

The upper breast is white and variably spotted or finely barred, the rest of the underparts white and densely barred with grayish to rusty brown, the bars sometimes overlying a narrow brownish band on the lower breast. Legs and feet are thickly feathered.

The juvenal plumage is grayish on the head and brown on the back and has fewer white spots. The facial disk is mostly black, but there are conspicuous white eyebrows and a pair of white muttonchops under the eyes. The underparts are finely and indistinctly barred gray.

Of three subspecies, only one occurs in North America.

VOICE: The advertising song is a trilling, high-pitched series of whistles, "hoo-hoo-hoo-hoo" or "ulululu" (Voous 1989), also rendered as "tu-wita-wit, tiwita-tu-wita, wita, wita" (Cramp 1985). A variety of other calls, often having to do with nest defense, includes alarm calls resembling those of a hawk or falcon ("kee-kee-kee").

FLIGHT: The flight of the Northern Hawk Owl is, well, hawklike: direct and dashing or fast, agile, and maneuverable when darting between bushes and trees (Voous 1989). It is also said to soar (Thompson 1891). An immature male demonstrated a display flight, gliding on spread wings with raised head and then circling (Nero 1995). The tail is uniquely wedge shaped in flight.

Fig. 107. The Northern Hawk Owl most commonly assumes a slanting posture, rather than perching upright like most other owls.

Fig. 108. The pointed wings and long tail lend the Northern Hawk Owl a distinctly falconlike appearance.

DAILY ACTIVITY PATTERN AND FEEDING: Primarily diurnal and apparently chiefly a sight hunter, the Northern Hawk Owl usually perches on treetops, roosting at night in a tree's canopy, not in cavities (Cramp 1985); but it hunts at night as well as by day. When actively looking for prey, this owl changes its usual diagonal position to a nearly horizontal one and may pump its tail up and down. Typically, it darts at prey in low flight but also hovers and makes foraging flights; it even catches prey in the air occasionally (Dement'ev and Gladkov 1966) or by running after it and can locate and catch prey hidden under 30 cm (1 ft) of snow (Duncan and Duncan 1998).

Whereas voles of various species are the main prey, other mammals caught range from shrews *(Sorex)* to hares *(Lepus)* and Ermine *(Mustela erminea)*. Birds (often caught in winter when rodents are scarce) range in size from White-crowned Sparrow *(Zonotrichia leucophrys)* to Spruce Grouse *(Falcipennis canadensis)*. Even carrion is taken on occasion (Duncan and Duncan 1998). One study in the Yukon revealed that voles of the genus *Microtus* were preferred while *Clethrionomys* were avoided, and the owls did not respond numerically to population peaks of the latter but did so to the combined densities of *Microtus* and the cyclic Snowshoe Hare *(Lepus americanus)* (Rohner et al. 1995).

While the Northern Hawk Owl caches food as reserves, particularly during nesting season (and especially after the young leave the nest), it does so in winter as well, stashing prey in crevices, on stumps and other places, and at the base of trees, sometimes pushing snow with its beak over the item to hide it (Duncan and Duncan 1998).

REPRODUCTION: Considering their chilly habitat, Northern Hawk Owls begin their breeding season remarkably early, in February, although they may start as late as June (Duncan and Duncan 1998). The male attracts (with singing) the female to a potential nest site, typically decayed hollows at the tops of broken trees or where branches have fallen off, old woodpecker cavities, and burned out stumps, or, occasionally, old stick nests. The sites are from about 2.5 to 5 m (8 to 16 ft) above ground.

The female lays from three to 13 (typically seven) eggs, which she incubates, starting with the first egg, for about three and one-half to over four weeks. She eats the pellets and, probably, droppings of the young to promote nest sanitation; meanwhile, the male has been providing for her from the onset of incubation and for the young. The chicks fledge when they are three to five weeks old, with the youngest leaving last, and are slow to acquire flight. They reach independence when they are about two and one-half to three months old (Duncan and Duncan 1998).

DISTRIBUTION AND HABITAT: An owl of the boreal forest, this species breeds in a broad circumpolar band from Scandinavia (where this bird has been extensively studied) through Siberia and northern Mongolia and the North American continent from Alaska to Labrador and Newfoundland. Northern Hawk Owls nest rarely as far south as northern Minnesota, northern Wisconsin, and in Montana, where, in 1990, 1994, and 1995, breeding

was confirmed, the first records for the western United States. Generally, this sedentary species is quite prepared to tough out a northern winter. However, in irruption years, dozens or even hundreds of these owls show up as far south as Illinois and Nebraska, and, in the West, Wyoming, Idaho, and northern Oregon, movements probably brought on by fluctuations in prey availability. In Montana, this species is normally seen from November to February and in May and June. In Russia and Fennoscandia, banded owls dispersed more than 1,000 km (600 mi) (Glutz von Blotzheim and Bauer 1980).

The breeding habitat is typical boreal forest of moderately dense conifers such as tamarack *(Larix)* and spruce *(Picea)* or where such trees are mixed with deciduous broadleaved trees—aspens *(Populus tremuloides)*, for example. Nests are close to open areas in the form of marshes or land cleared by logging or by fire (Duncan and Duncan 1998). A Canadian study detected nests only in postfire forest and not in nearby unburned coniferous forest; burned forests were only suitable for eight years after the fire, the optimum being two years (Hannah and Hoyt 2004). Open areas are essential in that they provide sufficiently high prey density for breeding owls.

SIMILAR SPECIES: No other owl resembles the Northern Hawk Owl nor behaves as it does; it is truly unmistakable. In flight it could conceivably be mistaken for a short-winged hawk (*Accipiter,* such as a Cooper's Hawk *[A. cooperii]*), but it has a wedge-shaped tail, quite unlike that of hawks, and a much larger head.

STATUS: Because of its remote habitat, the status of the Northern Hawk Owl in North America is difficult to assess, although this population is thought to be stable (Duncan and Harris 1997). Habitat degradation has resulted from fire suppression, and forestry practices (such as creation of mosaics of clear-cuts, forest remnants, and tree plantations) have likely reduced nest site and hunting perch availability while increasing prey (vole) populations (Duncan and Harris 1997). Modified clear-cutting could enhance Northern Hawk Owl habitat (Duncan and Duncan 1998). The owl is a Montana state species of concern.

REMARKS: Most Northern Hawk Owls live in a land of few people and have failed so far to acquire a useful fear of humans. Easily spotted and approached, some northern shooters call them "practice owls" (Austen et al. 1994). They may fearlessly attack humans approaching their nests and can inflict serious injuries (Lane and Duncan 1987). Native people eat them.

This owl has been used for falconry in Scandinavia.

NORTHERN PYGMY-OWL *Glaucidium gnoma*

Pl. 9

FIGS.: 8, 20, 24, 33, 44, 45, 52, 61, 78, 109–112

IDENTIFICATION: A diminutive, plump owl, about the size of a starling, with relatively small head and eyes, the latter yellow. There are two conspicuous blackish, white-rimmed ovals on the nape. The long, prominently barred tail is often jerked from side to side or cocked.

The dome-shaped head and the back are brown, reddish brown, or grayish (distinctly grayer in the Rocky Mountains), with numerous white to buff spots, especially on the head. The nape bears a pair of large rhomboid or oval black spots rimmed in white, which vaguely suggest eyes. The brown facial disk is poorly developed and has white or buff spots arranged in concentric semicircles, and the eyebrows are white, as is a narrow bib under the yellowish bill.

A broad reddish brown band, also with buffy spots and lighter than the head, spans the upper breast and continues down the sides toward the flanks. This horseshoe pattern frames a contrasting white lower breast, belly, and lower flanks, all of which are boldly streaked with black brown. The tail has broad black brown bands, separated by narrow white ones. The legs are feathered, the large and sturdy feet sparsely so. Some individuals have been designated rufous morphs (forms) or gray morphs.

The juvenile Northern Pygmy-Owl is similar to the adult, but most of the head is grayer and has diffuse, indistinct spots, with the crown and back unspotted.

There are three subspecies recognized in the United States, two of which occur in California: *Glaucidium gnoma grinnelli*,

Fig. 109. The disheveled appearance of this just-fledged Northern Pygmy-Owl illustrates the difficulties of living in a very small cavity with siblings and numerous leaky corpses brought in by the parents. Note the out-sized feet.

which lives in the western parts of the states from California into British Columbia (the latter boasting an additional race on Vancouver Island), and the race *G. g. californicum,* which is found generally east and in all the western states.

Because of geographic differences in their vocalizations (and a few other features), certain populations of the Northern Pygmy-Owl may represent different species, according to some researchers (Howell and Webb 1995).

VOICE: This owl is most vocal in the hour after sunset and within the first two hours after sunrise (Noble 1990). The most familiar call is a monotonous "toot-toot-toot" series, given at one- to two-second intervals (hence slower than a Northern Saw-whet Owl's call, and it is also lower in pitch), which serves as the advertising song and may at times be embellished with trills. It can be heard year-round, and at times, each toot may have a slight downward inflection, audible if the listener is sufficiently close. Pair members may duet, or vocalize antiphonally; the female's pitch is distinctly higher (Holt and Petersen 2000). There is also a call that resembles a short version of the Western Screech-Owl's *(Megascops kennicottii)* bouncing ball song. The "weet" call, given with a rising inflection by adults, appears to be a warning call to alert present pair members or young to avian predators.

The begging call of the young is said to be reminiscent of a katydid's song (Holt and Petersen 2000) but to my ear, from a distance of 15 m (50 ft) or so, strongly resembles the sequestration (or contact) call of the Bushtit *(Psaltriparus minimus)* given in a long, persistent series.

FLIGHT: The flight of this species is surprisingly noisy, particularly for an owl, and resembles that of a woodpecker or a shrike: it is undulating, with bouts of flapping alternating with gliding with wings folded. It can also resemble a bat as it darts through the crown of a bare tree. The rather pointed (for an owl) wings are short, and the tail is long, traits that are often observed in flight.

DAILY ACTIVITY PATTERN AND FEEDING: Small and generally secretive, this owl is easily overlooked, particularly when it is sitting still and not being mobbed—it is a frequent target of passerines, because it forages by day and perhaps into the night. Swift in flight, it is gone in an instant, defying identification and even notice, unless the observer happens to live in an area where the species is seen with some regularity.

Fig. 110. A female Northern Pygmy-Owl has just retrieved a Western Fence Lizard *(Sceloporus occidentalis)* that she had cached 24 hours earlier in a creek bed.

Its daily activity pattern is not well known. For example, it is not known whether it roosts in cavities like other small owls often do, although it has been seen to roost in conifers (Bull et al. 1987). Although usually solitary during the nonbreeding period, occasionally two Northern Pygmy-Owls can be seen near one another at that time.

The Northern Pygmy-Owl bathes in birdbaths, pools of standing water, and creeks, and it enjoys sitting in the sun as well.

A hunting Northern Pygmy-Owl is alert, craning its neck and head and weaving like a boxer as it seeks to get a better look at and assess distance to potential quarry. It frequently jerks its tail from side to side in an odd stiff fashion, or bobs it up and down. Like other owls, it drops on prey from a perch, or approaches it in a zigzag pattern before plunging down. The Northern Pygmy-Owl is not above pulling birds from nest cavities (Holt and Petersen 2000); one attempted to look into a small nest box in October. In spring it may sit high in a tree, perhaps to locate bird nests to raid. It appears to have favorite perches.

This species eats insects including beetles and crickets, reptiles such as various lizard species, birds ranging in size from hummingbirds to California Quail *(Callipepla californica),* and mammals, chiefly rodents including voles *(Microtus),* pocket gophers *(Thomomys),* and tree squirrels *(Tamiasciurus).* One owl returned to an Alameda County yard twice weekly in October and decimated a population of Western Fence Lizards *(Sceloporus occidentalis).* It seemed to ignore various birds visiting a feeder. Another pursued a fly-fisherman's lure and was caught.

The Northern Pygmy-Owl is famous for tackling very large prey and not shying away from a fierce wrestling match, which usually ends in the owl's favor. The predator may be so engrossed in its endeavor that it allows itself to be picked up together with its prey. Having overcome and killed its quarry, the owl usually removes the head if it is a vertebrate and eats the carcass in pieces, typically in two installments separated by a few hours. One study showed that females ate more small mammals than did males, which took more birds (Earhart and Johnson 1970).

REPRODUCTION: Compared to other owls, the courtship of the Northern Pygmy-Owl is poorly known. Males seem to attract females with their advertising (tooting) song, sometimes given from the opening of a nest cavity. The responding female may solicit the male with tooting, trilling, or whistling (behavior normally thought to be typically male), but at least one tooter was observed being mounted (Holt and Petersen 2000).

In early April 2006, one male persistently sang from a branch of a western sycamore *(Platanus racemosa)* in Alameda County, California, about 2.5 m (8 ft) from what turned out to be the nest cavity, a vacant woodpecker hole 13.5 m (44.5 ft) up in an adjacent sycamore (apparently higher than previously reported in the literature). Shortly, an owl thought to be the female disappeared behind the trunk of the tree, perhaps inspecting the nest hole. A minute or so later, she flew off swiftly, producing a loud screech resembling the noise made by fighting cats or raccoons. The male now shifted his perch to a tangle of branches, approximately 3.6 m (12 ft) from and slightly higher than the nest hole, where he was presently joined by the female. The two birds copulated for about 20 seconds, the male rapidly beating its wings; the female, and then the male, flew off.

The use of old woodpecker cavities for nests is typical. There is conflicting information whether these are modified in any way by the addition of feathers, bark strips, or moss as substrates.

The female lays from two to seven eggs in mid-April, with clutches of three to four the most common. Incubation takes about four weeks, with the male bringing food to the female in the cavity (which she may leave, responding to his call, or which he may or may not enter to make the delivery). The female resumes hunting for herself and to contribute to the raising of the young when these are about nine days old (Holt and Petersen 2000). In mid-May, the Alameda County female mentioned above was seen to cache a freshly caught Western Fence Lizard on a rock in a creekbed and retrieve it almost exactly 24 hours later, at which time she ate the front half and cached the remainder in a clump of sprouting leaves at the end of a sycamore branch. Two days later, it was determined that the female was incubating four eggs, together with two European Starling *(Sturnus vulgaris)* eggs, almost certainly deposited before she took over the cavity. Both European Starlings and Violet-green Swallows *(Tachycineta thalassina)* were repeatedly seen peering at length into the nest hole while the owl sat on her eggs.

The young fledge at about three weeks of age, sometimes virtually synchronously at three to five minute intervals or within one to two days, and are feeble, generally clumsy fliers that usually stay together, perched not too far from the nest. Both parents

Fig. 111. Exiting its nest hole in a western sycamore, a diminutive Northern Pygmy-Owl with a gleaming white breast is easily overlooked.

bring food at this stage for several weeks. It is not known when the young gain independence.

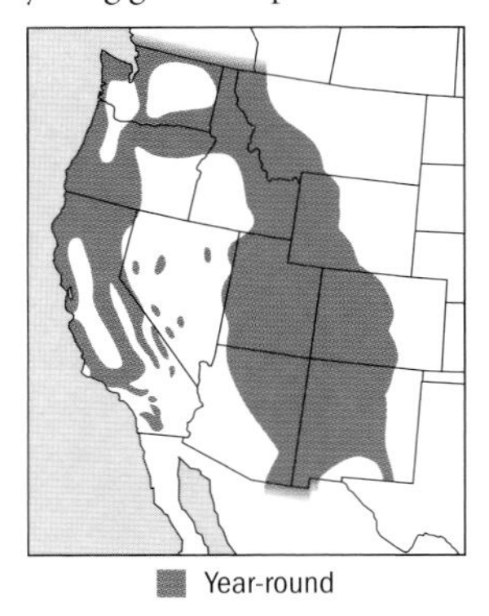

DISTRIBUTION AND HABITAT: In the United States and Canada, the Northern Pygmy-Owl is a distinctly western species, ranging generally from north-central British Columbia and southwestern Alberta south through the mountainous parts of the Rocky Mountain states and the ranges of the Great Basin to Arizona and New Mexico, and through the Pacific states south through most of Mexico (including the tip of Baja California) to Guatemala and Honduras.

In Oregon, the Northern Pygmy-Owl was most common in the western Cascades and least so in the Willamette Valley and the high deserts, with the other mountain ranges falling in between (Sater et al. 2006).

In California, this species occurs as a breeder the length of the Sierra Nevada, the Warner Mountains, the Cascade Range, and down the Coast Ranges and the Tehachapi Mountains and some of the southern California mountain ranges, such as the San Gabriel and San Bernardino Mountains. It is absent from the Central Valley, our Great Basin areas, the southernmost mountains, and the deserts.

Although it seems tied to coniferous forest and in fact breeds in forests of pines and firs, this owl also nests in oak-laurel and other oak woodlands, in riparian habitats, and in hardwood-conifer forests. In the course of Spotted Owl *(Strix occidentalis)* surveys in the northwestern counties of California, Northern Pygmy-Owls were detected with surprising frequency. A study of the Northern Pygmy-Owl in fragmented forests on the Olympic Peninsula in Washington State determined that it clearly preferred structurally diverse and older forests for nesting and foraging, and forest openings and stands of saplings were used least for foraging and not at all for nesting (Giese and Forsman 2003). In Oregon, the number of Northern Pygmy-Owls detected per search increased with the average diameter ranking of the dominant overstory trees, and the greatest numbers occurred in Douglas-fir and ponderosa pine forests (Sater et al. 2006).

SIMILAR SPECIES: The Elf Owl *(Micrathene whitneyi)* is short-tailed and prefers different habitats than the Northern Pygmy-Owl; it is even smaller and has diffuse grayish ocher ventral steaks. The Flammulated Owl *(Otus flammeolus)* sometimes shares its habitat with the Northern Pygmy-Owl but is dark eyed and has a well-developed facial disk and small but usually visible ear tufts.

In poor light and seen in silhouette flying high in a tree, a Northern Pygmy-Owl can be mistaken for just another starling because of its small size, fairly pointed wings, and rapid wing-beats.

An often look-alike species, the Neotropical Ferruginous Pygmy-Owl *(G. brasilianum)* occurs in Arizona; it has a very different voice and lives in the Sonoran Desert with much vegetation, habitat no Northern Pygmy-Owl would seek out. (It is also found in Texas.)

STATUS: Although presumed nowhere common, the Northern Pygmy-Owl is more numerous than generally believed and widespread in suitable habitat in California. Breeding Bird Survey trend assessments (from California data collected from 1967 to the present) (Sauer et al. 2005) are ambiguous, which may not only simply reflect the imprecision of the survey methodology but also the difficulty of locating this species. During the 1980 Christmas Bird Count, 27 were counted at Año Nuevo (San Mateo County).

Clear-cutting of commercially valuable timberlands obviously is detrimental to this species, given its habitat preferences. It is suspected stable in Wyoming but is listed as a state species of special concern. It is designated a stewardship bird species in Nevada.

REMARKS: Although it lacks obvious ear tufts, the Northern Pygmy-Owl, having spotted a potential predator, creates these projections on the spot by selectively raising certain groups of unspecialized feathers on its head while the rest lie flat. Simultaneously, it erects its white eyebrows and rictal bristles, along with the feathers of the lower "cheeks," and assumes the typical small-owl Count Dracula pose (see "Going Undetected" in the section "An Owl's Life"). All of these efforts, combined with its coloration and color patterns, serve to make the owl resemble a branch stub, producing a concealment posture that was also observed in wild

Fig. 112. Most owls are masters at hiding, and this Northern Pygmy-Owl is no exception.

pygmy-owls trying to hide from mobbing songbirds (Holt et al. 1990).

High on any owler's hope-to-see list, the Northern Pygmy-Owl generally turns up when least expected. Although the chorus of warning calls of outraged songbirds as they mob the little fellow alerts the curious naturalist, most often the owl's simple and easily recognized song gives away its presence.

FERRUGINOUS PYGMY-OWL *Glaucidium brasilianum*

Pl. 10

FIG.: 113

IDENTIFICATION: A diminutive, plump owl about the size of a starling, with a relatively long tail (for an owl), broadly striped underparts, and a pair of conspicuous blackish oval spots on the nape that are bordered by a narrow white margin and vaguely resemble eyes. The actual eyes are bright yellow. The tail is usually (though not always) boldly barred and is often twitched from side to side or cocked.

The brown facial disk is small and not well developed except for the inner margins above the beak, which form conspicuous white eyebrows. White spots flank the throat, and a larger white spot is centered between two dark patches on the upper breast. The lower breast and belly are adorned with about six more or less continuous dark brown to reddish brown streaks, and wide dark brown side panels connect the flanks to the upper breast.

The Ferruginous Pygmy-Owl is polymorphic; that is, it comes in a variety of colors. The form most commonly seen in the United States is grayish brown on the head and back; there are fine whitish streaks on the crown and nape, and white spots of varying sizes appear on the scapulars, back, and upperwing. The blackish brown to grayish brown tail has about six narrow white bars, or it may have seven or eight well-defined rufous bars, about equal in width to the dark brown bars separating them; individuals with faintly barred or even unbarred tails also occur.

The juvenal plumage is like that of the adult except that there are few or no whitish streaks on the head.

There are up to 15 subspecies of this owl over its entire range, with perhaps four occurring in North America. A rufous form of this owl is apparently rare in the United States, but many individuals are intermediates between it and the grayish brown morph. South of the U.S. border, Ferruginous Pygmy-Owls have narrow whitish tailbars instead of rufous ones. Subspecific characteristics and range of variation within a population are difficult to sort out.

VOICE: The most commonly heard vocalization is the advertising song, which consists of a series of single, upward-inflected toots that are given in a series of 10 to 30 notes for five minutes and as long as five hours. Easily imitated by simple whistling, the more serious owl mimic can reproduce the tone by blowing over the top of an 80 percent full 20 oz. (590 ml) soda bottle. Other vocalizations include chitters (given by females and young when begging) and two-note alarm calls (Proudfoot and Johnson 2000).

FLIGHT: Like other owls, Ferruginous Pygmies drop off a branch, glide 30 to 70 m (100 to 230 ft), and then rise steeply to another perch. Powered flights are typically straight and short, with shallow undulations.

DAILY ACTIVITY PATTERN AND FEEDING: Although the Ferruginous Pygmy-Owl is quite diurnal, most calling occurs at dawn and dusk. This owl also flies and to some extent calls at night as well. Hunting, too, is weighted toward the crepuscular hours, though foraging continues throughout the day if the owl is hungry or has a mate and young to feed, and the bird will readily respond to mimicked calls all day, especially during nesting season.

Like the Northern Pygmy-Owl *(G. gnoma)*, the Ferruginous starts its hunts from a perch and peers into nest cavities searching for nestlings. Insects are a major food item, chiefly grasshoppers and crickets but also moths and lightning bugs. Amphibians, reptiles (such as lizards), and rodents up to rat *(Sigmodon* and *Dipodomys)* size are prey, along with birds, including Eastern Meadowlark *(Sturnella magna)* (Proudfoot and Beasom 1997) and Gambel's Quail *(Callipepla gambelii)* (Proudfoot and Johnson 2000), species larger than the owl.

REPRODUCTION: Male Ferruginous Pygmy-Owls in Texas defended territories as large as 600 m (2,000 ft) across year-round (Proudfoot and Beasom 1996), whereas in Arizona these areas measured only about 160 m (525 ft) across (Hensely 1954). The territorial advertising song serves in courtship, with the male also proffering food to the female in typical owl swain fashion. Monogamous, the pair may, however, not necessarily consort with one another throughout the year. Copulation may be enhanced by allopreening before and after.

Each territory holds several nest cavities, the active site typically near two others that may be used to cache prey, for occasional roosting by the male, or as an alternate site should predators destroy the first clutch. Usually 4 to 6 m (13 to 20 ft) above ground (but as high as 12 m [40 ft]), the nest cavity is typically an old woodpecker hole (such as those made by Gilded Flickers *[Colaptes chrysoides]*) located in mesquite, cottonwood, or ash

(Prosopis, Populus, Fraxinus) and additionally in saguaro *(Cereus gigantea)* in Arizona and live oak *(Quercus)* in Texas. Other natural cavities may also be used, even tree forks and a hole in a sand bank (Johnsgard 2002). Nest boxes are welcomed, too. Ferruginous Pygmy-Owls are quite prepared to tolerate people and have nested within less than 10 m (33 ft) of human residences (Proudfoot and Johnson 2000).

The female incubates her two to seven (most commonly three or four) eggs for about four weeks, the male acting as the sole food provider until the young are three weeks old. They fledge about a month after hatching and depend on their parents for food for another two months or so.

DISTRIBUTION AND HABITAT: The Ferruginous Pygmy-Owl ranges from southernmost Texas and southern Arizona through Mexico and Central America to Tierra del Fuego at the tip of South America.

In Texas, this owl is found in undisturbed live oak–mesquite forest as well as in mesquite brush, ebony *(Pithecellobium ebano)*, and riparian areas. Arizona birds typically live at elevations from 300 to 1,300 m (1,000 to 4,300 ft), in

Fig. 113. The Ferruginous Pygmy-Owl, scarce in Arizona, seems common in tropical Mexico and Central America. Note the heavy streaking on the breast and belly.

riparian formations of cottonwood and mesquite or in upland paloverde-mesquite-acacia associations and saguaro-paloverde country. Historically, this owl was far more widespread in Texas and Arizona, and in Mexico and other Latin countries, it can be found in a wide variety of habitats, including backyards in agricultural areas.

SIMILAR SPECIES: The very similar Northern Pygmy-Owl's range overlaps that of the Ferruginous, but the two species occupy very different habitats, and the Northern Pygmy has a spotted, not streaked, crown and a blackish tail with white bars versus the rusty, wide bars (in the United States) of the Ferruginous' tail. The sympatric Elf Owl *(Micrathene whitneyi)* is smaller yet, has diffuse marking (instead of clean streaks) on the underparts, and has a very short tail. Other smallish owls in the Ferruginous Pygmy-Owl's range are substantially larger and have conspicuous ear tufts.

STATUS: Habitat destruction and alteration in Texas and Arizona have severely limited this owl's distribution, in stark contrast to its apparent abundance in Mexico, for example. The owl may have disappeared from much (if not most) of its range in Arizona, where areas were designated as critical habitat following the species' listing as federally endangered by the U.S. Fish and Wildlife Service in 1997; it was subsequently delisted federally but remains state endangered in Arizona. It is listed as threatened by the state of Texas. This species may be more sensitive than other small owls to habitat changes in the small areas where it occurs in the United States, because it is on the edge of its range (Mehlman 1997).

REMARKS: My first acquaintance with this active little owl occurred in Mexico, where one responded to my whistle and perched no more than 10 feet in front of me, calling back. Within a minute, a second owl appeared and, taking advantage of an opportunity where it saw one, copulated with the first arrival.

ELF OWL *Micrathene whitneyi*

Pls. 11, 21

FIGS.: 42, 57, 114

IDENTIFICATION: A tiny owl, smaller than a starling, with a short tail, no ear tufts, and yellow eyes.

The head and upperparts are gray and brown with rusty, reddish ocher to buff irregular spots. The facial disk is moderately developed and is rusty to reddish ocher, with small gray patches to the outer sides of the eyes, conspicuous white eyebrows, and a prominent white triangle or small crescent at the disk's margin under each eye, edged in black.

Fig. 114. An Elf Owl's size is appropriate for preying on arthropods.

Bright white spots form a continuous outer margin on the scapulars and run down the edge of the folded wing. The underparts are whitish, with indistinct broad vertical streaks that are gray, reddish ocher, and finely barred with black brown. They form a continuous band across the upper breast. The legs and feet are sparsely feathered.

The juvenile is overall gray with much reduced and paler buff spots. The white marks below the eyes are triangular and conspicuous. The gray underparts are very indistinctly marked with narrow dark transverse bars.

VOICE: The advertising song is the so-called Chatter Song, composed of five to seven notes. It resembles a puppy's yipping and serves to advertise the presence of a male to females and to proclaim and defend a nest site. It can also be varied in length. Other calls include the female's "sheee" call, given while copulating, and an alarm call that is a single sharp bark (which, in an agitated owl, is accompanied by side-to-side tail flicks). The Elf Owl sings more often when there is plenty of moonlight. Like other owls, it vocalizes most at dusk and before sunrise.

FLIGHT: This owl flies with quick wingbeats in direct pursuit of prey. It also hovers over crawling arthropods before dropping down on them. As an insect eater, its flight need not be silent; it is not.

DAILY ACTIVITY PATTERN AND FEEDING: The Elf Owl roosts in dense trees that are often, though not always, evergreen, or in clumps of mistletoe. It also roosts in old woodpecker cavities, sometimes as pairs. These different sites are used as needed for thermoregulation; apparently this owl has problems keeping its body temperature down when there is low ambient humidity (as in deserts) (Ligon 1968).

The Elf Owl forages from perches and in flight, catching insects in the air, on the ground, and on vegetation. It takes advantage of nocturnal arthropod attractants: campfires, outdoor lights and such, flowers, and hummingbird feeders. Insects may be beaten out of vegetation or extracted from flowers and leaves as the owl hangs from the plant (Henry and Gehlbach 1999). It is most active at dusk and before dawn (Ligon 1968).

Large insects are favored, such as crickets, scarab beetles, and hawkmoths. Other arthropods are taken, too, including solpugids, centipedes, and scorpions (which are apparently disarmed by removal of the stinger or the terminal body segments [Ligon 1968]). A few vertebrates are taken, for example lizards, small snakes, and young kangaroo rats *(Dipodomys)*. Prey items, especially large ones, are cached in nest holes (Ligon 1968).

REPRODUCTION: The male Elf Owl defends a territory, and in one study, there was no overlap of the home ranges of nine radio-tagged individuals (Gamel and Brush 2001). As in other species, the male takes the female on invitational tours of potential nest cavities, singing his Chatter Song and diminishing the volume as he backs into the cavity, seeking to lure her in (Ligon 1968). He also presents her with food.

Except for nest boxes, which are also accepted, woodpecker holes provide the nest sites. Such holes need not be in trees; one famous nest in southern Arizona was in a telephone pole. Even woodpecker holes in fence posts are acceptable, as well as such cavities in yucca *(Yucca)* and agave *(Agave)* stalks. There is no nest per se; indeed, insisting on Spartan quarters, one owl spent parts of three nights pulling out an old grass nest, the handiwork of a House Sparrow *(Passer domesticus)* (Henry and Gehlbach 1999). Although nests may be no more than 10 m (30 ft) apart (Goad and Mannan 1987), they are normally much more than that.

Usually the female lays three eggs in late April or early May, although from one to four are possible, and incubation lasts from three to about three and a half weeks. She may feign death when handled by humans (Ligon 1968), as do some other owls. The young, poor flyers early on, leave the nest when they are about four weeks old, coaxed by their parents.

An Elf Owl may utilize the same tree with other hole-nesters if it has multiple cavities; not only songbirds such as flycatchers *(Myiarchus* and *Myiodynastes)* but also even screech-owls *(Megascops)* and pygmy-owls *(Glaucidium)* may share the same habitat and compete for holes.

DISTRIBUTION AND HABITAT: The Elf Owl breeds from the southern half of Arizona, southeastern and southwestern New Mexico, and the southwestern and southern tips of Texas south through most of Mexico to Puebla, and at the tip of the Baja California peninsula.

In California, where the continued occurrence of the Elf Owl is questionable, the species was found in the extreme southeastern corner of the state, from the Colorado River west to Joshua Tree National Monument (Riverside County), north to a little beyond Needles (San Bernardino

County), south to the Chuckwalla Mountains (Riverside County), and, along the river, to Winterhaven (Imperial County).

Whereas the Elf Owl was traditionally thought to be mainly associated with saguaro cactus *(Cereus giganteus)* desert, this owl is in fact more numerous in other habitats—for example, in chaparral in Texas (Gamel and Brush 2001), canyon riparian woodlands of sycamore *(Platanus)*, cottonwood *(Populus)*, walnut *(Juglans)*, ash *(Fraxinus)*, and box-elder *(Acer)* in Arizona and New Mexico, and in oak and oak-pine forests on mountain slopes in Arizona. In that state, it may be the most abundant raptor in upland deserts and is found to almost 2,000 m (6,500 ft) in mountain canyons (Henry and Gehlbach 1999).

Above all, the owl, as a secondary cavity nester, requires the presence of old woodpecker holes. In California, such habitats included saguaros (before they were for all practical purposes extirpated in the state), riparian woodland of paloverde *(Cercidium)*, willow *(Salix)*, and mesquite *(Prosopis)*, and desert oases such as Corn Springs and Cottonwood Springs (both in Riverside County), where cottonwoods with holes are often available. The relentless spread of tamarisk *(Tamarix)* has all but destroyed the riparian habitat where this owl was most common in California and southern Nevada.

Elf Owls, if they still occur in California, migrate, like their kind in other states, to Mexico in early August, returning mid-March and April (Garrett and Dunn 1981).

SIMILAR SPECIES: No other owl is as small as the Elf Owl. Screech-owls occurring in the same habitat are not only substantially larger but also have conspicuous ear tufts. In some areas, the ranges of the two species of pygmy-owls overlap that of the Elf; however, pygmy-owls are bigger and have long tails, in contrast to the Elf Owl's short appendage, and their ventral streaking is narrow, crisp, and well defined, not wide, blotchy, and diffuse.

STATUS: The Elf Owl is considered extirpated or nearly so in California. (See "Owl Conservation" in the section "Owls and Humans.") Its status in the rest of its range is not well known.

It is California state listed as endangered.

REMARKS: This tiny owl in parts of its range manages to survive in habitats it shares with other small species, namely the Flammulated Owl *(Otus flammeolus)*, screech-owls *(Megascops)*, and the scrappy pygmy-owls *(Glaucidium)*, all of them potential competitors for nesting cavities.

Like Eastern Screech-Owls *(Megascops asio)*, Elf Owls, in areas outside California, are unwitting hosts in their nests to blind snakes *(Leptotyphlops)*, originally brought in as food, that survive and go on to provide a pest control service by eating ant larvae and pupae and fly maggots that would otherwise take advantage of the food cached for the owlets' use. Certain ants may also live as commensals or mutualists with Elf Owls without attacking the young (Henry and Gehlbach 1999).

BURROWING OWL *Athene cunicularia*

Pls. 12, 21

FIGS.: 3, 32, 34–36, 39, 49, 62, 67, 75, 76, 85, 86, 115–117

IDENTIFICATION: A small, brown, yellow-eyed owl, about pigeon size, with long legs and a dome-shaped (as opposed to ball-shaped) head that lacks ear tufts. It sits on the ground or on exposed (often low) perches. The sexes are similar; however, the female is a bit smaller and darker and often more heavily marked on the underparts, with a dirty-looking facial disk.

The head and upperparts (including the folded wings) are brown, speckled and spotted with buff white. The tan facial disk is vertically flattened and relatively small, shaped a bit like a pair of goggles, with usually prominent white or off-white brows and rictal fans. A conspicuous large white rhomboid chin patch below the yellow gray beak spans the lower margin of the disk and is in turn bordered by a dark brown band.

A wider dark brown band with large white spots stretches across the upper breast and borders the white throat patch. The rest of the underparts are cream to white, with widely spaced brown bars and chevrons, except for the unmarked lower belly, legs, and undertail coverts.

The short, broadly banded tail is usually mostly hidden by the wing tips. The legs are incompletely feathered, and the toes are nearly bare.

The juvenile Burrowing Owl has a dark brown head, with a facial disk somewhat darker than that of the adult, with buffy brows and whitish rictal fans. The head and upperparts are brown, sometimes indistinctly spotted, and a large oblong buff patch is

Fig. 115. Burrowing Owls may fly about by day. Note the long wings and short tail.

conspicuous on the folded wing, which is otherwise marked like an adult's. The upper breast bears a dark brown band, darkest where it borders the throat patch. The remainder of the underparts is buff to ocher and completely unmarked.

VOICE: The advertising song of the Burrowing Owl is sung by the male only and can be rendered as "coo coooo." It is reminiscent of the song of a dove or of a cuckoo clock but is a bit raspy and can be mimicked by whistling the two syllables while forming the letter K in your mouth. Other calls, charmingly named, are the copulatory Smack Call, given by the female; the Coo or Coo Coo Song, variant of the advertising song, given during copulation by the male, who may end the entertainment with a Tweeter Call; and finally a Male Warble, of which there is also a female version. Screams, clucks, and chatters are directed at predators, perceived and real, near the nest (Haug et al. 1993).

The young beg with a rasping call, which famously mimics a rattlesnake's *(Crotalus)* buzz and probably deters some predators from entering a burrow.

FLIGHT: Disturbed, this owl flies off in long, shallow undulations. It can also hover and pursue insects on the wing but is not as buoyant in flight as a Short-eared Owl *(Asio flammeus)*, nor does it stay aloft as long as the larger species often does.

DAILY ACTIVITY PATTERN AND FEEDING: The Burrowing Owl is active any time of day or night, though it usually spends daylight hours roosting near or in the mouth of its burrow or a ground hollow unless it has young to feed. Outside the nesting season, it may either live singly or in pairs but almost invariably at a burrow that is the center of activity as well as an attractant to edible arthropods (Coulombe 1971), which seek shelter in this environment offering relatively high humidity and darkness.

The birds spend the day desultorily preening and stretching, and pairs roosting together occasionally allopreen, more so as breeding ramps up. Dust bathing helps to remove parasites, and rain makes the owls run about excitedly and stretch and flap their wings (Haug et al. 1993). The Burrowing Owl also sunbathes, lying down with fully spread wings and with feet projecting to the rear (see fig. 36).

A tilted-up head slowly moving in an arc means the owl is tracking a potentially dangerous raptor's passage across the sky, and an unfamiliar visitor is greeted with an alert, upright stance and a series of curtsies.

This owl was long thought to be a chiefly diurnal hunter, but a study in Colorado showed that most foraging activity takes place at dusk and throughout the night (Pezzolesi and Lutz 1994). Traps baited with mice have shown that in winter, Burrowing Owls do not approach the bait before late dusk and are therefore strictly nocturnal, and pre-independent juveniles at the nest burrow become noticeably more active at dusk. Adults with young, however, forage even in the middle of the day (J. Barclay, pers. comm. 2006). Prey is pursued from a perch, but also from a hover and on foot, when the owl runs or even hops after the quarry.

The Burrowing Owl feeds on a variety of invertebrates such as scorpions, centipedes, beetles, earwigs, and crickets, and on small mammals, particularly voles *(Microtus)*, and, in deserts, on heteromyid rodents. Oddly, in Alberta, Canada, these owls ate 1.45 voles for every mouse they consumed, notwithstanding that trapping of these small animals in the same area showed a 113:6 ratio of mice to voles (Schmutz et al. 1991) (however, voles are notoriously much more difficult to trap than mice). Birds are also captured, principally small passerines such as Horned Larks *(Eremophila alpestris)*, although one owl snatched the last in a string of Mallard *(Anas platyrhynchos)* ducklings (S. Ferreira, pers. comm. 2005).

Every fall, several presumably dispersing young Burrowing Owls appear on the Farallon Islands, 45 km (28 mi) west of the Golden Gate, and feed on the peaking population of the non-native House Mouse *(Mus musculus)*. When this food source crashes in early winter, most of the owls leave or starve, but a few stay on and switch to preying on Ashy Storm-Petrels *(Oceanodroma homochroa)*—a seabird with its own survival problems—which they catch at night. Gulls returning to the islands to breed in spring take a dim view of the predators and kill them with pecks to the head (J. Buffa, pers. comm. 2006).

Crustaceans, amphibians, and reptiles (including turtles) round out the menu. In one study in central California, mammals formed the most common pellet item except between December and February (Thomsen 1971). Surprisingly, in 18 Burrowing Owl pellets collected in Bakersfield (Kern County), the dominant vertebrate prey was Mexican Free-tailed Bats *(Tadarida brasiliensis)*; the same study found accumulations of Western Toad *(Bufo boreas)* carcasses at other owl burrows in nearby areas (Hoetker and Gobalet 1999).

REPRODUCTION: Although the Burrowing Owl is territorial in the immediate vicinity of the burrow, this species is semicolonial, with nest sites sometimes only 30 or so feet apart. Usually monogamous, males on occasion have been noted with more than one female. Nonmigratory pairs remain together year-round, and migrants may arrive on the breeding grounds singly or in pairs (Haug et al. 1993).

A courting male flies upward to about 30 m (100 ft), hovers for a few seconds, drops halfway down, and rises again to repeat the display (Grant 1965). Allopreening, food presentation, and singing are other components of male persuasion. Standing tall, ruffling his body feathers and flashing his white markings, the male prepares to copulate with the female, who also flashes her white but stands less tall, without raising her body feathers. Romantic throughout, the male sings while he mates with her, and, at the end, he may warble and tweeter (Haug et al. 1993).

For nesting, the Burrowing Owl makes use of the holes of various animals such as ground squirrels *(Spermophilus)* (the most common sites in California), prairie dogs *(Cynomys)*, badgers *(Taxidea)*, and other mammals, as well as tortoise holes, erosion-caused cavities, and natural rock cavities. In Florida, three pairs actually tried nesting in scrapes on lawns but failed (Cavanagh

Fig. 116. A tug-of-war results when a juvenile Burrowing Owl tries to take a cricket from its father.

1990). When burrows are unavailable, the owls may dig their own, where possible (such as in sandy soils), with both pair members participating, kicking dirt backward with their feet and excavating with their beaks, or they can modify existing holes. A pair of owls can move a substantial amount of soil.

The nest cavity is lined with dried, shredded cow or horse dung. The burrow's entrance, usually on some sort of slope or rise, which perhaps provides a lookout and coincidentally reduces the chance of flooding, is also adorned with dung—in some places, at least, to attract edible insects (see "Hunting and Eating" in "An Owl's Life")—and, often, with found objects. Such decorating peaks during incubation (J. Barclay, pers. comm. 2007).

In the West, these owls mostly lay eggs from late March to late April, the latter in California (Thomsen 1971). However, clutches have been found in California as late as June in the Imperial Valley and August in Orange County (P. Bloom, pers. comm. 2006). There may be up to 11 eggs, but from four to eight is most common. Such reproductive enthusiasm reflects the high mortality rate. Incubation lasts four weeks or a little more and is done by the female, the male supplying her with food in early morning and in the evening (Zarn 1974).

The young owls pop out of the nest burrows at about two weeks of age but quickly retreat when danger threatens, especially when alerted by the adults' warning calls. In the open, they engage in all manner of interesting behavior (see "Finding and Watching Owls"), including capturing prey, as they learn how to function as proper little owls.

During the day, adults perch with the youngsters at the burrow mouth or nearby on fence posts, telephone wires, and other elevated sites, where they apparently act as sentries, giving alarm calls at the approach of a predator. They also carry on limited foraging, supplying the young with food—during the day chiefly insects, the dead and dying bodies providing pouncing practice for the youngsters (Green 1983). Nearby satellite burrows provide emergency shelters to both adults and young, and the nest burrow is used for caching food (Haug 1985).

At the age of about one and a half months, the young leave the parental burrow but still remain nearby and begin to join the adults in crepuscular hunting flights.

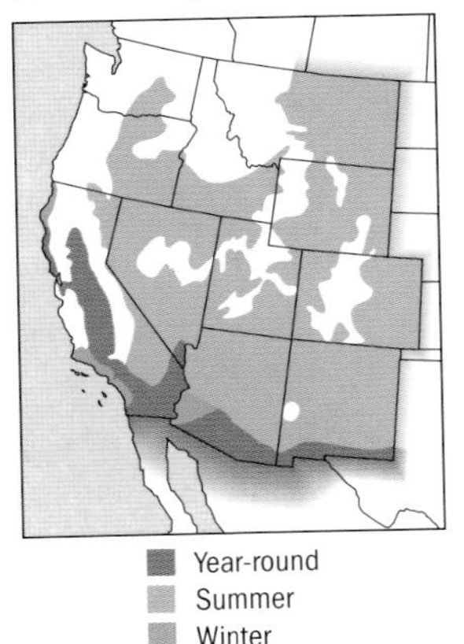

DISTRIBUTION AND HABITAT: The Burrowing Owl is, for the most part, a central United States and western species. Apart from long-isolated populations in Florida and on a handful of Caribbean islands, the owl inhabits a broad swath of land running from the southern parts of the central and western Canadian provinces south through the prairie states to the Gulf of Mexico coast in Texas, through the rocky Mountains and drier parts of the west coast states into Mexico (including Baja California), and south through Central America and South America (except the Amazon Basin) to Tierra del Fuego. As many as 18 subspecies are recognized, with *A. c. hypugaea* extending into the West.

In California, this species breeds in the Great Basin and Modoc Plateau areas of the northeast (whence it migrates south or southwest in winter), the southern part of the Sacramento Valley, locally in the greater San Francisco Bay Area south to Santa Barbara County, on some of the Channel Islands, and in the San Joaquin and Salinas Valleys, and in scattered localities of some southwestern counties. Relatively large populations can be found in eastern Alameda County, Santa Clara County, and western San Joaquin County, in the Central Valley, and in the Imperial and Palo Verde valleys (Imperial County). The Carrizo Plain (San Luis Obispo County) is also believed to have many Burrowing Owls.

Dispersing young may simply move into satellite burrows or travel over considerable distances. In California, Burrowing Owls drift around frequently in winter (J. Barclay, pers. comm. 2007). In a study of this species across the northern Great Plains (including parts of Canada), stable isotopes extracted from the feathers of nestlings were compared with the isotopes in feathers of post-hatch-year owls breeding in the sampled areas the following year to determine the natal origin of the breeders. Although some owls returned to the area of their birth, others commonly appeared to relocate to new breeding sites hundreds of kilometers away, including some that moved over 1,000 km (600 mi). Almost half of the owls examined in Montana probably came from Canada; overall, 20 to 25 percent of young owls in the northern Great Plains permanently moved away (Duxbury et al. 2003). Very limited banding returns show that owls wintering in California come from Idaho, Washington, and Oregon.

Although the relentless march of development threatens this species' habitat in much of California, it so far has been sufficiently tolerant of humans to hang on in commercial areas, on golf courses and airports, in parks and sewage plants, and in other similarly "improved" modern landscapes. Sadly, the early stages of property development create attractive (but short-term) habitat for the owls.

When not forced to make do with such artificial habitats, Burrowing Owls are inhabitants of open country, specifically deserts, shrub-steppe, and grassland or prairie that has been grazed to just the right height. Interestingly, in a South Dakota study, densities of this owl were higher on sites grazed by Buffalo *(Bison bison)* than on cattle-grazed and ungrazed grounds (Murray 2005). Wyoming Burrowing Owls preferred Black-tailed Prairie Dog *(Cynomys ludovicianus)* colonies over those of White-tailed Prairie Dogs *(C. leucurus)* because they were more open and had shorter vegetation (Martin 1983). The presence of some sort of hole, preferably one of a large rodent with satellite burrows, is a great attractant. Additional desirable features include nearby elevated perch sites such as fence posts or utility poles, phone cables

Fig. 117. If food supply and nesting habitat are adequate, Burrowing Owls do not hesitate to breed near busy roads and highways.

or wires, and, often, proximity of a road, the latter perhaps inviting because careless arthropods and mice are so easily spotted and caught as they cross it.

SIMILAR SPECIES: No other owl that even vaguely resembles the Burrowing Owl is found in its habitat or abroad by day. Other eartuftless owls of that size are denizens of forests and have different proportions and color patterns. The other open-country owl sometimes active during the day, the Short-eared Owl, is also yellow eyed but is twice as large, is streaked instead of barred, and has short legs and tiny ear tufts.

STATUS: Although this species is clearly declining in the West, its situation is not desperate in California. However, it has been mostly extirpated in nearly all the southwestern counties of the state, or their populations are so fragmented that long-term survival there is questionable (Kidd et al., in press). The owl does well in some strongholds notwithstanding habitat destruction, predation, pesticide poisoning, and being buried alive by agricultural machinery and getting chopped up by wind turbines. It is a California species of special concern.

Populations are thought to be stable in Wyoming, but the owl is a state species of special concern. Montana and Utah list it as a species of concern, Colorado designates it as state threatened, and in Idaho the breeding population is considered imperiled.

REMARKS: This owl is commonly called the Ground Owl. Other local names include Billy Owl, Prairie Owl (in the prairie states), and Howdy Owl (in Florida), from the owl's habit of alighting with a bob and curtsy. It is also called the Prairie Dog Owl, for its close association with that rodent in states where they both occur. With prairie dog populations in decline, fewer burrows are available to the owl; old burrows in abandoned prairie dog towns will not do, because the services of the mammals are required to keep the grass down.

The Burrowing Owl is more resistant to carbon dioxide than other birds, presumably because it nests in burrows (Haug et al. 1993).

It is beset by a number of parasites, especially several species of fleas, one of them the Human Flea *(Pulex irritans)*, which may outnumber all the others (Smith and Belthoff 2001a).

A Burrowing Owl appeared on a ship at sea, 71 km (44 mi) off the Monterey peninsula, and welcomed aboard another traveler, a California Towhee *(Pipilo crissalis)*, which it eventually ate (D. Shearwater, pers. comm. 2005).

SPOTTED OWL *Strix occidentalis*

Pls. 13, 20

FIGS.: 11, 43, 118

IDENTIFICATION: A midsized, plump brown owl, spotted whitish both above and below, with a large round head that lacks ear tufts and dark brown eyes that appear black. The female is larger than the male.

The pale brown facial disk is marked with indistinct concentric rings and is shaped like the cut half of an apple, with the right half arching noticeably higher above the eye than the left. The eyebrows and rictal fans are white to off-white. The beak is greenish yellow.

The chocolate, brown, or chestnut brown head and upperparts bear round to oval whitish spots. The upper breast is colored and marked like the upperparts, but the belly and flanks are patterned with dark brown connecting arrowhead-shaped crossbars bisected by wide streaks, making the parts appear checkered or, alternately, white spotted. The wingtips envelop the short tail, which has broad crossbars. The feathered legs and feet are buff, with dark curved marks.

The juvenile Spotted Owl is buff to pale brown, with indistinct darker bars. Because the contour feathers have not yet grown (except for the upperparts), the facial disk is conspicuously small.

Two races occur in California, the Northern Spotted Owl *(Strix occidentalis caurina)*, of the Pacific Northwest, and the California Spotted Owl *(S. o. occidentalis)*, the Sierran and southern Californian race, which is a bit smaller and paler, with slightly larger spots. A third race, the still paler Mexican Spotted Owl *(S. o. lucida)*, which does not occur in California, may in fact be a separate species (Barrowclough and Gutiérrez 1990).

VOICE: The advertising song, called the Four-note Location Call, is a series of four hoots: "hoo—hoo-hoo—hooo" given by both sexes to announce the territory. Other calls include the Series Location Call, essentially a seven- to 14-series variant of the FLC, as well as whistles, barks, chittering, and coos. The fledged young beg with a raspy whistle: "sweeet!" (Forsman et al. 1984).

FLIGHT: Although not very fast, this owl is an agile and highly maneuverable flyer, alternating gliding with fast wingbeats.

DAILY ACTIVITY PATTERN AND FEEDING: The Spotted Owl roosts by day under the forest canopy at variable heights, moving to cooler or less sunny sites to avoid heat stress, to which it is very susceptible because of its thick plumage, which is more typical of more northern owls (Barrows 1981). It spends much time preening (and allopreening with its mate when they are perched together). Primarily a nocturnal hunter, it forages diurnally only opportunistically (Sovern et al. 1994) but may begin its night hunts as much as one hour before sunset. A still-hunter of mostly rodents, the owl also hawks insects and bats from a perch, catching them in flight.

Food choices depend on what is locally available. Whereas the Northern Spotted Owl feeds mainly on Northern Flying Squirrels *(Glaucomys sabrinus)* outside of California, within the state Dusky-footed Wood Rats *(Neotoma fuscipes)*, preferentially selected by the owl (Ward et al. 1998), make up the bulk of the diet. The Spotted Owl can be found in younger forests because wood rats *(Neotoma)* abound here (Folliard et al. 2000). Somewhat surprisingly, weasels *(Mustela)*, Spotted Skunks *(Spilogale putorius)*, and Ringtails *(Bassariscus astutus)* have been recorded as prey (Forsman et al. 2004). Other foods include voles *(Phenacomys,*

Fig. 118. A Northern Spotted Owl in a coast redwood forest.

Arborimus, and *Clethrionomys),* mice *(Peromyscus),* pocket gophers *(Thomomys),* rabbits *(Lepus* and *Sylvilagus),* various birds (a large proportion of them smaller owls), amphibians, reptiles, and invertebrates, including snails (Forsman et al. 2004). The California Spotted Owl in addition feeds on various diurnal squirrels *(Sciurus, Tamiasciurus, Spermophilus,* and *Eutamias)* and moles *(Scapanus).*

Like many other owls, when there is an excess of food, the Spotted Owl may cache it on stumps, under logs, and in other hiding places, in warm weather selecting sites cooler than the surrounding areas (Gutiérrez et al. 1995).

REPRODUCTION: Although the home range of a pair of Spotted Owls may overlap with that of neighbors, the birds are territorial and defend the greater nesting area (Solis and Gutiérrez 1990). Monogamous birds, these owls breed irregularly. Male and female begin to roost together in February and early March, preparatory to selecting a nest site. Nesting habitat usually consists of dense, multilayered mature forest with a closed canopy, if available; however, nesting occurs in many other places. The nest itself may be the top of a broken-off tree, a tree cavity, a clump of dwarf mistletoe *(Arceuthobium),* an abandoned raptor or squirrel nest, or accumulations of leaves and branches in tree forks. Rock ledges or cavities are also used, but rarely in California. There are local differences in housing tastes. Beyond scraping out a depression and adding a few feathers, no construction per se takes place.

Eggs are laid from early March to mid-April. The Spotted Owl has a low reproductive rate, laying most commonly one or two and rarely three or even four eggs per nesting attempt, and raising only one brood. Incubation lasts four and a half weeks or a little less.

The young acquire their second down (the juvenal plumage) after 10 to 20 days, at which point the female no longer broods them continuously and begins to forage again, a task up until then left to the male. A study of the diet and fledging success of breeding Northern Spotted Owls showed that pairs that successfully fledged young consumed significantly larger prey than pairs that were unsuccessful breeders (White 1996).

The young prefledge about five weeks after hatching, though they may accidentally fall out of the nest before then. Either way, they are days from flying and must scramble up trees and such nearby, though the premature tumblers may initially spend much more time on the ground. The parents look after them for another two to three months, as late as early fall.

Dispersal is encouraged by the adults' withholding food, and though the young are awkward, they begin to capture prey as they prepare to leave their natal areas. The adults expand their winter home ranges after the young leave to seek suitable habitat (Gutiérrez 1996). One Northern Spotted Owl was discovered roosting in early October on a steel rail just inside an old military bunker on the Marin Headlands north of San Francisco (far from its normal habitat of coastal forest) and stayed all day, to the delight of hundreds of birders (A. Fish, pers. comm. 2006).

DISTRIBUTION AND HABITAT: The geographic range of the three races of Spotted Owl is western North America and Mexico. The Northern Spotted Owl is found from southwestern British Columbia through western Washington and Oregon and parts of northern California. The California Spotted Owl occurs from northern California south into Baja California. The Mexican Spotted Owl occurs from southern Utah and southern Colorado south through eastern Arizona, western and central New Mexico, part of western Texas into Mexico down the length of the Sierra Madre Occidental and parts of central Mexico.

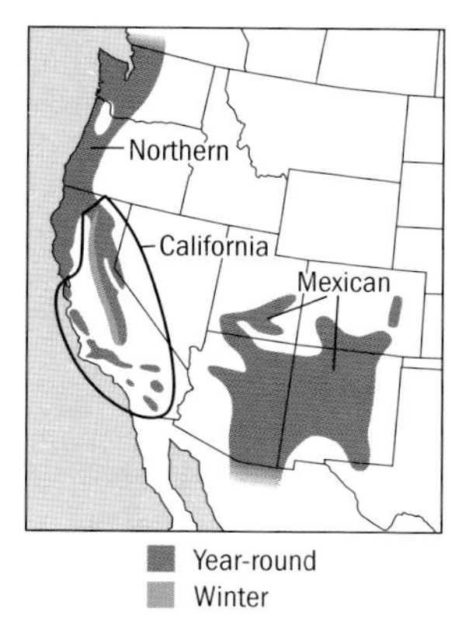

In California, the Northern Spotted Owl occurs in the northern and central Coast Ranges, as far south as Marin County and east to the relatively dry Klamath Mountains and eastern edge of the Cascade Ranges and the Central Valley, from near sea level to about 2,150 m (7,000 ft) (Gutiérrez 1996), occupying various types of chiefly coniferous forests, including coast redwood *(Sequoia sempervirens)*, Douglas-fir *(Pseudotsuga menziesii)*, grand fir *(Abies grandis)*, ponderosa pine *(Pinus ponderosa)*, and Shasta red fir *(Abies magnifica shastensis)*, as well as mixed conifer-hardwood and mixed evergreen forests. At the southern end of its range, it may even use pure hardwood forests that have year-round water (Gutiérrez 1996). In a given forest, the owls select specifically parts that have a more complex vegetation structure with larger trees and a more complete canopy than is generally the case for the rest of the forest. Such habitats serve for roosting, nesting, and foraging, though the owls also hunt in more open areas with smaller trees.

The California Spotted Owl occurs from the southern Cascades down the length of the western Sierra Nevada (and, locally, east of the crest), west through the Tehachapi Mountains, and in the Coast Ranges from Monterey County to San Diego County as well as in the Transverse and Peninsular Ranges. It seeks out similar habitats (see above) in the western Sierra Nevada and on insular mountains and ranges in southern California. In the Sierras, mixed conifers are the norm, but in southern Californian mountains, nest productivity was shown to be greatest in lower elevation live oak/bigcone Douglas-fir *(Quercus, Pseudotsuga macrocarpa)* habitat (LaHaye et al. 1997), but nesting also occurs in mixed conifer and in riparian hardwood habitats. The latter, found for example in the deep canyons of the Los Padres National Forest, is composed of coast and canyon live oaks *(Quercus agrifolia* and *Q. chrysolepis)*, western sycamore *(Platanus racemosa)*, white alder *(Alnus rhombifolia)*, California bay *(Umbellularia californica)*, and cottonwood *(Populus)* (Verner et al. 1992). There are historic records of these owls nesting in San Diego County at Oceanside and San Onofre. In the Santa Lucia Mountains, the California Spotted Owl is rare and local; most are found in the dense canyons of the Ventana Wilderness (Monterey County) (Roberson 2002).

The majority of Sierran nests are found at elevations from 900 to 2,100 m (3,000 to 7,000 ft), with some down to at least 300 m (1,000 ft) (or perhaps lower) or as high as 2,360 m (7,740 ft) (Verner et al. 1992). Sierran owls move downslope in winter, establishing home ranges below snow level (Laymon 1989).

Roost-site studies of the Mexican Spotted Owl in Arizona and New Mexico suggested that mixed-conifer forests provide stable and favorable conditions year-round, in contrast to pine-oak forests, which force the owl to make seasonal adjustments in roost-site use (Ganey et al. 2000).

SIMILAR SPECIES: The only owl that can be mistaken for a Spotted Owl is the quite similar appearing Barred Owl *(Strix varia)*, a fairly recent immigrant to California but long known from other western states to the north. It also has a big, round, tuftless head and dumpy body but is a bit larger and, often, paler brown dorsally. Importantly, the head, upperparts, and breast are distinctly barred, and the belly is streaked. The Spotted Owl, by contrast, has a spotted breast and a checkered belly, as well as a spotted head and back. Barred Owl × Spotted Owl hybrids incorporate traits of both species, including some vocalizations.

STATUS: The status of the Spotted Owl is still the subject of much debate, although more is known about its numbers and distribution than about any other owl on earth. There have been complaints from some quarters that the amount spent on Spotted Owl research takes away funds for the study of other imperiled species, but in fact, Spotted Owl research is the study of an entire threatened ecosystem and so benefits many species.

The Northern Spotted Owl is federally listed as threatened in California, Oregon, and Washington (the U.S. Fish and Wildlife Service also considers it threatened in British Columbia), but that race has no California state regulatory designation; it is state listed as threatened in Oregon and endangered in Washington. The California Spotted Owl is listed as a California species of special concern (federal listing was denied), and Nevada designates it a species of conservation priority. The Mexican Spotted Owl is federally listed as threatened in Colorado, New Mexico,

Texas, and Utah (and is also considered by the U.S. Fish and Wildlife Service to be threatened in Mexico); it is state listed in Arizona and Colorado as threatened.

REMARKS: Besides other owls, the Northern Goshawk *(Accipiter gentilis)* also preys on the Spotted Owl, especially juveniles. Fishers *(Martes pennanti)* and Pine Martens *(M. americana)* may also eat eggs and incubating females (Gutiérrez et al. 1995).

Around humans, these owls are disarmingly tame and often approach closely to imitations of their song and displays of artificial mice, practices that are to be discouraged because they disrupt normal activity. Spotted Owls are the spirits of the ancient forest and are to be respected.

BARRED OWL *Strix varia*

Pls. 14, 20

FIG.: 119

IDENTIFICATION: A midsized to large, plump, brown owl with a very big head lacking ear tufts, the barred breast contrasting with the streaked belly, and with dark eyes. The female is larger than the male, sometimes substantially so.

The head and back are transversely barred with brown to gray brown on a buff or white background, and this pattern continues onto the upper breast so that the owl appears to be wearing a head scarf and muffler. The facial disk is pale grayish tan with two to three variably distinct concentric rings around the eyes and a conspicuous dark brown margin that also forms frown lines. The brows and rictal fans are whitish.

The outer scapulars are spotted, as are some of the upperwing coverts. The tail is broadly barred dark brown. The lower breast (differing sharply from the upper breast), belly, and flanks are boldly streaked, providing a diagnostic contrast on the underparts unique to the species. The legs are buffy to white with irregular bars or spots, the toes mostly bare. The yellow beak is large and projects well above the white throat.

The juvenile, at fledging, has typically ragged-looking plumage, which is buffy, the feathers having broad white tips and indistinct narrow or broad barring on the head and underparts (chiefly on the breast). The upperparts largely resemble those of the adult, and the facial disk is whitish to pale gray tan.

VOICE: The advertising song, "hoo'-hoo-to-hoo'-ooo," or "hoo-hoo-hoo-to-whooo'-ooo" (Johnsgard 2002), has been often transcribed as "who cooks for you, who cooks for you all?"

An ascending series of six to nine hoots, ending on a descending "hoo-aw" (which may be given separately) is, like the previous call, given by both partners of a pair (McGarigal and Fraser 1985). Dueting pairs produce a hair-raising assortment of screams, gurgles, yells, and cackles, with maniacal laughter thrown in for good measure, leading a listener to believe that Walpurgis Night is in full swing; it is guaranteed to stir some life into a tired, late-returning hiker.

The young's begging call is a hissing whistle, audible for long distances.

FLIGHT: This owl adeptly maneuvers between trees, preferring open forest, where it uses specific flyways (Nicholls and Warner 1972). It flies buoyantly with slow wingbeats.

DAILY ACTIVITY PATTERN AND FEEDING: The Barred Owl spends the day roosting, using tree cavities or branches, typically next to the trunk. Temperature sensitive, it seeks out roosts under a closed canopy. It is chiefly a crepuscular and nocturnal forager, launching itself from perches at prey below.

A great variety of animals is taken, among them mice and voles and other mammals, birds up to the size of a Ruffed Grouse *(Bonasa umbellus),* amphibians (which may be hunted on foot) and reptiles, and numerous invertebrates, including crayfish, which, along with fish, may be caught by wading in shallow water (Mazur and James 2000). The owl may also catch fish by dropping down from a perch over water (Smith et al. 1983). It is the scourge of screech-owls *(Megascops)* but takes larger owls as well (including at least one Great Gray *[Strix nebulosa]*), though it in turn is eaten by Great Horned Owls *(Bubo virginianus).* Prey not consumed on the spot may be cached on branches, on snags, or in a nest.

REPRODUCTION: The Barred Owl defends territories year-round and has a permanent pair-bond, although outside the breeding season, the pair members roost separately. Both male and female increase the frequency of calling as courtship begins. Deciduous trees are often favored for nesting, probably because of the greater abundance of cavities formed by disease or by fallen or broken

Fig. 119. A fledgling Barred Owl, like most other owls, can swallow remarkably large rodents whole.

branches. This owl also uses the broken tops of trees as well as nests built by crows, ravens, hawks, and squirrels. It also takes advantage of nest boxes.

A clutch of two to three eggs, laid from March through April, is typical for the Barred Owl, though there may be as many as five. Incubation takes about four to four and half weeks. For the first two weeks, the hatchlings are brooded by the female, who does very little foraging during that period, with the male provisioning her. Both parents supply the young with food after that. The young prefledge four to five weeks after hatching and, having fallen to the ground, climb leaning trees (Mazur and James 2000). At the age of 10 weeks, they begin to fly, and toward fall, as the parents stop supplying food, they disperse.

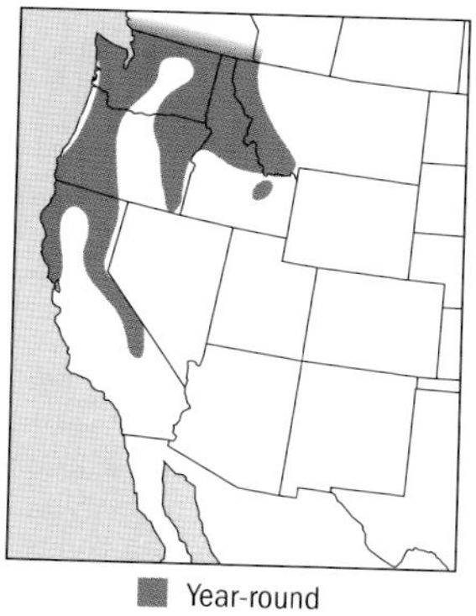

DISTRIBUTION AND HABITAT: The Barred Owl is found in the east from south of James Bay in Canada down the eastern and central United States nearly to the Mexico border, skirting the central and southern Great Plains. A disjunct population, forming a separate subspecies, occurs on the central Mexican plateau.

Within historic times, this species has spread across the northern Great Plains across

Canada in the wake of tree planting and the establishment of riparian forests and shelter belts that provided Barred Owl habitat across the midwestern United States and southern Canada. The mosaic of habitats created by logging and forest fragmentation first encountered by the owls in the Rocky Mountains served their needs, permitting them to establish themselves from southern Alaska, most of British Columbia, and most of Alberta south through western Wyoming, most of Idaho, nearly all of Washington, and the forested parts of Oregon into California.

After its arrival in California in 1981, the Barred Owl spread rapidly (Dark et al. 1998). As of 2006, it is found from the northern border down the North Coast Ranges to Marin County, across the Cascades and down the length of the Sierra Nevada. It is numerous in northern California but more scattered in the Sierras. A typically sedentary species, its advance is the result of dispersal, not of migration.

In California, this owl is at home in disturbed and fragmented forests, along with old-growth forests and riparian habitat. It likes river/bottomland but also readily takes to coastal redwood *(Sequoia sempervirens)* and Douglas-fir *(Pseudotsuga menziesii)* forest as well as conifer–tan oak *(Lithocarpus densiflorus)* associations, and it is found in redwood–white alder *(Alnus rhombifolia)* riparian zones. Although the Barred Owl is usually thought of as an inhabitant of river valleys, it is in fact also found as high as 1,500 to 1,800 m (5,000 to 6,000 ft) in red fir–white fir *(Abies magnifica* and *A. concolor)* forests. It may very well live in other types of forests and woodland as well but has gone undetected in the absence of Spotted Owl *(S. occidentalis)* searches (using taped songs in such habitats).

SIMILAR SPECIES: The closely related Spotted Owl could be mistaken for a Barred Owl but has spotted and checkered underparts and a spotted head, versus the Barred Owl's transverse barring on the breast, bold streaks on the belly, and clearly barred head. No other dark-eyed, medium-sized owl without ear tufts and with a streaked belly is found in its habitats.

STATUS: The Barred Owl is a bird on the move. As it spreads southward in California, it displaces the smaller Spotted Owl, usurping its habitats. It is conceivable that this species will invade other ecologic communities that offer nest sites and food, such as certain oak woodlands and riparian woodlands, already in scarce supply in the state. Certainly it seems at present that Barred Owls can look after themselves and are sufficiently versatile in their habitat and diet that the species can be expected to become a permanent component of the California avifauna unless, to protect Spotted Owls, it is somehow prevented from doing so.

REMARKS: In California, Barred Owl habitat so far differs somewhat from that used by populations in the eastern United States, where the owl has an affinity for old mixed forests of hardwoods and conifers without an understory (but near swamps and riparian zones). In the east, nests are often found near those of Red-shouldered Hawks *(Buteo lineatus)*, with which the owl often shares much the same habitat (and prey, taken at different times of the day), and in fact the two species may use the same nest alternately (Bent 1961).

Although the Barred Owl is an intriguing addition to California's bird community, its effect on the native Spotted Owl is worrisome. In addition, because of its wide-spectrum diet, this owl could also have an impact on other raptors, including, for example, the Western Screech-Owl *(Megascops kennicottii)*, which would be directly affected by becoming meals of a predator to which it is not accustomed. Of equal concern is this owl's taste for amphibians, a number of which are threatened or endangered in California.

The Barred Owl can be fierce in its nest defense and has on occasion attacked people that ventured too close to the site.

GREAT GRAY OWL *Strix nebulosa*

Pls. 15, 20

FIGS.: 9, 13, 120

IDENTIFICATION: A very large owl with a usually startled expression, appearing bigger than a Red-tailed Hawk because of its enormous head and thickly layered plumage. In fact, this owl weighs considerably less than the smaller-looking Great Horned Owl *(Bubo virginianus).* It has a very striking dark-rimmed facial disk that suggests a microwave antenna or radar dish, with concentric rings and with small-looking yellow eyes. The female is considerably larger than the male.

Upperparts are grayish brown to maroon brown with darker blotches, irregular bars, and paler gray or dirty white vermiculations. The beak and cere are dull to bright yellow. Below the facial disk, two bright white crescent moons form a conspicuous necklace divided by a black spot. Underparts are grayish white, with diffuse large variably dark brown blotches on the breast and smaller ones on the belly along with indistinct barring. Undertail coverts have wide pale brown transverse bands. The tail, long for an owl, is broadly barred except for the central rectrices (which are more mottled). Legs and toes are densely feathered except for the soles.

The juvenile is indistinctly barred pale brown and dirty white on the underparts, more broadly and cleanly so above. The developing disk starts as crescents on both sides of the face, forming a conspicuous shield of feathers on the forehead that makes the young owl appear to be frowning.

VOICE: The advertising song of the adult Great Gray Owl, most often given after one o'clock in the morning, is a repeated series of deep, booming hoots, evenly spaced. However, researchers find it very difficult to get this species to call (R. Montgomery, pers. comm. 2006). The male's voice is lower than the female's. Another call, given perhaps for contact or territorial defense, consists of low double hoots. Chittering calls and single hoots, bleating, and nickering are also included in the vocabulary. A nasal "ee wheet" is a begging call given by females during courtship and the nesting

Fig. 120. Juvenile Great Gray Owls can look quite hawklike before the enormous facial disk develops.

cycle, and a higher-pitched version is used by the young to beg (Beck and Winter 2000).

FLIGHT: The flight of the Great Gray Owl, as that of most other owls, is direct. It has very large, broad-handed wings, which span up to 1.5 m (5 ft). Gliding alternates with flapping, the wingbeats deep and slow, somewhat like those of a heron, but this species is also capable of speed. Because of its small body, the wing-loading is very low, which facilitates maneuvering in dense forests, as well as hovering.

DAILY ACTIVITY PATTERN AND FEEDING: This owl typically roosts by day in the lower part of the canopy near the trunk of a usually tall coniferous tree, but it can also be observed foraging in daylight. Preferring to hunt at dusk, it is equally likely to be seen foraging in early morning and in late afternoon in winter, particularly on overcast days. Nocturnal hunting occurs chiefly when there are young to feed, and in winter (Bull and Duncan 1993), in addition to foraging by day.

The Great Gray Owl is a forest bird, and its rounded wings seem adapted to hunting amid trees, but this species is rarely far from open ground such as meadows and pastures, where it finds its rodent prey. Unlike a Red-tailed Hawk *(Buteo jamaicensis)*, which scans an entire meadow, the Great Gray concentrates on a small area for a period of time, presumably because it is listening rather than only looking for prey, and then moves on to another site if unsuccessful.

Hunting flights begin from a perch (still-hunting) and, if the quarry is hidden, terminate in hovering before the owl plunges vertically down onto the prey. The Great Gray Owl has acute eyesight but also such exceptional hearing that it can, by sound alone, locate prey under a half meter (1.5 ft) of snow, catching it sight unseen with its feet by diving through crust or powder, sometimes face first (Nero 1980). It can break through crust hard enough to support the weight of a man (Holt et al. 1999).

The diet, at least in North America, is almost exclusively composed of little rodents, surprising for such a large owl. Voles (chiefly *Microtus*) and pocket gophers *(Thomomys)* are by far the most common prey in the western United States, including the Yosemite area. Other small mammals appear also in the diet, such as deer mice *(Peromyscus)*, flying squirrels *(Glaucomys)*, moles *(Scapanus* and *Condylura)*, and shrews *(Sorex* and *Blarina)*. Very few birds, up to the size of a grouse, are eaten, along with occasional frogs and beetles.

REPRODUCTION: The Great Gray Owl seems to breed very irregularly, perhaps at intervals of three or four years. Prey abundance appears to be a major governing factor (Winter 1985).

The owls may form pairs as early as January but often much later. The male seeks to impress a female by vigorously plunging into snow as if he were hunting. As in other owls, courtship feeding and allopreening are important features. Copulation takes place after the male chases the female.

Two kinds of nest sites are used by this owl. The majority outside of California are squirrel nests and old stick nests of diurnal raptors, such as the Northern Goshawk *(Accipiter gentilis)* or Osprey *(Pandion haliaetus)*, and of the Common Raven *(Corvus corax)* and other sizeable birds. In California, the Great Gray seems to greatly prefer the broken tops of large conifers, always near meadows. In Yosemite National Park, for example, trunks of nest trees averaged 114 cm (45 in.) in diameter at breast height, with nests on average 14 m (45 ft) above ground (Beck and Winter 2000), and this site type is used elsewhere as well. Some artificial broken tops made with chainsaws readily attracted occupants in California (Beck and Smith 1987), and nests made of wire and sticks installed in Canada were promptly used.

The eggs, which are more oval than those of most hole-nesting owls, are laid at one- to three-day intervals (resulting in young of staggered ages and sizes) between March (late March in California) and May, with heavy snow postponing laying (Bull and Duncan 1993). Clutches number two to five eggs, with two or three more common (but up to nine in the Eurasian race).

Incubation lasts four weeks or a bit more, with the female receiving all her food from the male, delivered at dusk and at night.

The young grow rapidly, provided food is plentiful, and the female initially leaves them alone for only very brief periods. She goes off to forage and roost on her own when they are about two and a half to three weeks old. In Oregon, females abandon young at the age of three to six weeks, with the male continuing to feed the offspring for up to three months.

At the age of three to four weeks, the young fall or jump from the nest, well before they can fly, and scramble up windfalls, fallen branches, saplings, and such to get above the forest floor. They become independent at the age of about three months and forage on their own. Dispersing from their natal area, they seldom move far, with most traveling less than 50 km (30 mi). They most commonly breed at the age of three years, occasionally earlier.

DISTRIBUTION AND HABITAT: The Great Gray Owl occurs across Eurasia in boreal and in dense coniferous forests as well as in the taiga, and it occupies similar habitat in North America, from central Alaska across much of Canada to the Great Lakes, with the southernmost populations reaching Colorado and the southern California Sierras. Great Grays are famously nomadic and, in years when small rodents in their boreal haunts fail, they (chiefly juveniles) move southward in numbers to southern Canada and the northern United States.

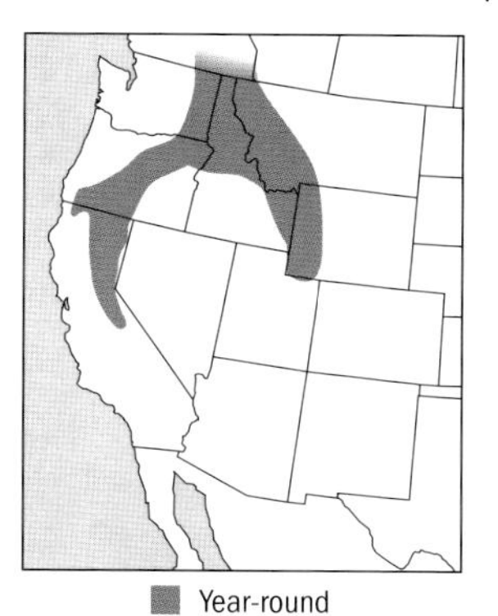

The most southerly Great Gray Owls in California (and the world) may be found in Sequoia National Forest (Tulare County) (Beck and Winter 2000), but the core nesting population is in the greater Yosemite area, nearly all of them west of the Sierra crest, although the owl likely nests on

Fig. 121. Great Gray Owl habitat, a meadow in Yosemite National Park. Owls are most often found in the narrow "fingers" of such meadows.

the east slope as well (Beck and Winter 2000). In Yosemite, the Great Gray Owl's distribution coincides with that of the red fir *(Abies magnifica)*. It has also been found in Fresno, Alpine, Nevada, Sierra, and Plumas counties, and spottily elsewhere in northern California (for example, the Warner Mountains) (G. Gould, pers. comm. 2006). It is very rare in the Cascades and Siskiyous (Beck and Winter 2000). However, a substantial breeding cluster exists just north of the California border in the northern Siskiyou Mountains in Oregon, near Ashland and Jacksonville, with some owls found at surprisingly low elevations (600 m [2,000 ft]) and even nesting in madrone-oak woodland. Individuals from that population not infrequently cross into California, especially in winter, and have been seen in northern counties such as Del Norte and Modoc (B. Woodbridge, pers. comm. 2006). One active nest presumably associated with that population was found in California (Fetz et al. 2003).

In the Sierras, Great Gray Owls inhabit both stringer meadows and large meadow complexes where, however, they seem to prefer the meadows' "fingers" and smaller side meadows, generally just wide enough for a person to throw a rock across and hit the opposite trees (R. Montgomery, pers. comm. 2006).

Some postbreeders, nonbreeders, and juveniles, at least in the

Yosemite Sierra, may move upslope to as high as 2,835 m (9,300 ft) (Gaines 1977).

In winter, northern Great Grays may move south, sometimes into very open areas. Yosemite birds and others may relocate downslope to snow-free places similar to their breeding habitats. In spring, they may then undertake exploratory upslope trips, only to come downslope again if the snow is still too deep, sometimes making the round trip in one day (R. Montgomery, pers. comm. 2006) using tree corridors (C. Stermer, pers. comm. 2006). Others remain on territory at or below the snow line (Beck and Winter 2000).

In California, Great Gray Owls occur in conifer-oak-woodland habitat composed of black oak *(Quercus kelloggii)*, incense-cedar *(Calocedrus decurrens)*, gray pine *(Pinus sabiniana)*, and ponderosa pine *(Pinus ponderosa)* at about 600 m (2,000 ft), and upslope from there, in mixed conifer forests of ponderosa pine, sugar pine *(Pinus lambertiana)*, Douglas-fir *(Pseudotsuga menziesii)*, and white fir *(Abies concolor)*, with a few oaks *(Quercus)*, up to about 2,000 m (6,500 ft). From 1,800 m to 2,700 m (6,000 to 9,000 ft), they live in red fir *(A. magnifica)* forests. The majority of nests are found at elevations ranging from 760 to 2,450 m (2,500 to 8,000 ft) (Beck and Winter 2000).

In Idaho and Wyoming, the great majority of Great Gray Owls is found in the lodgepole pine *(P. contorta)*, Douglas-fir, and aspen *(Populus tremuloides)* zone. In northeastern Oregon, nests were discovered in all forest types, whereas in central Oregon, the species is found in coniferous forests (Bull and Duncan 1993).

SIMILAR SPECIES: The Barred Owl *(S. varia)* and the Spotted Owl *(S. occidentalis)* are also big headed and hornless, but the Great Gray Owl is truly unmistakable. No other owl has the subtle fog gray and mahogany brown plumage or such a magnificent facial disk in which shine the small, yellow, staring eyes. No owl is as large (although two are heavier) as the Great Gray, nor as long tailed (except for the diminutive Northern Pygmy-Owl *[Glaucidium gnoma]*).

Just after the fledglings leave the nest, they can be confused with young Great Horned Owls *(Bubo virginianus)*; but at any age, the latter has a brown facial disk, whereas the Great Gray's is always gray (Beck and Winter 2000), and the young Great Horned already has ear tufts.

STATUS: California population numbers are uncertain, with estimates ranging from 40 to 250 individuals, although some estimates are probably unrealistically high. The state of California has designated the Great Gray Owl as endangered, and its continued existence in the state must be considered precarious. At present, this very modest population apparently is stable, but it would not take much to eliminate this species from the California breeding avifauna. Populations in Wyoming and Montana are also suspected to be stable, but both states designate the Great Gray a species of special concern.

REMARKS: The Great Gray Owl has an extraordinarily thick layering of feathers enveloping a not so large body and therefore tolerates snow very well, provided there is sufficient food. The soles of the feet are warm to the touch, unlike the cold feet of a duck, for example. In winter, however, many Great Grays outside California succumb to starvation in areas of heavy snowfall, although they can survive a 70 percent drop in body weight. At that point, they become sluggish and preternaturally tame around humans, with all their energy focused on finding food (Voous 1989). Innocent as they are of people, Great Grays are easily caught in winter by banders who drag a fake mouse (a cat's toy) by a string across the snow, then clap an angler's landing net over the pouncing raptor (Nero 1980).

A fierce defender of its young, the Great Gray can raise its rictal bristles to threateningly expose its beak, much like a snarling dog bares its fangs. The female is especially aggressive when the chicks are small. There are reports of people having lost eyes, or having wound up with broken bones (Mikkola 1983). Helmets and other protective gear for banding operations are recommended.

LONG-EARED OWL

Asio otus

Pls. 16, 20

FIGS.: 16, 22, 26, 122, 123

IDENTIFICATION: A midsized, often slender-appearing owl, about as big as a crow, with long and nearly always highly conspicuous ear tufts (laid flat in flight) that often appear placed closer together than those of other owls and contribute to this species' pointy-headed appearance. The sexes are generally alike in colors and patterns, with the female darker and rustier below.

The head is finely marked with gray, white, and rust, and the tufts are blackish with white and ocher or rust margins. The bright facial disk is buff or pale to dark rust with variable amounts of black surrounding the yellow to orange eyes. The eyebrows are tan to white, and a blackish frame surrounds the entire disk, bordered in turn by a narrow white and gray vermiculated rim.

The upperparts are finely marked and indistinctly barred with brown on a paler grayish white to pale ocher background, with a row of a few white spots running down the edge of the scapulars. The underparts are white to ocher, with diffuse blackish blotches on the throat and upper breast. The lower breast, flanks, and belly bear bold shaft streaks crossed by broad chevrons so that these underparts look somewhat checkered. The legs and toes are densely covered with pale to dark ocher feathers.

The juvenile has the head and body covered with whitish to pale tan down, with numerous narrow, pale gray bars. The ear tufts are grayish to tan down. The facial disk is blackish and rust, paler around the yellow eyes, and starkly contrasts with the rest of the plumage. Developing back and flight feathers are like those of the adult. By 10 weeks of age, the young look like adults.

VOICE: The male's advertising song consists of a series of about 30 (but as many as over 200) "hoo" notes, evenly spaced and audible for more than half a mile. The female gives a nest call, transcribed as "shoo-oogh," with a reedy quality, like a lamb's bleat (Marks et al. 1994). There are also barks, squeals, and mews.

Begging young produce piping calls, resembling a creaky gate or the creaking of wind-whipped trees (Marks et al. 1994).

FLIGHT: The Long-eared Owl is aerial and forages chiefly on the wing, the flight consisting of glides with intermittent deep wing-beats. It is adept in flying through thick cover, surprising for such a long-winged species.

DAILY ACTIVITY PATTERN AND FEEDING: Chiefly nocturnal, the Long-eared Owl roosts during the day in dense vegetation near open grounds. Roost sites include conifers, willows *(Salix)*, mesquite *(Prosopis)*, and tamarisks *(Tamarix)*. On migration, dozens of these owls may roost together, and in the breeding season, males may sometimes sit with others from nearby pairs while the females apparently sleep on their nests as they incubate (Marks et al. 1994).

Foraging may begin before sunset, the owl quartering over open ground a few feet up. In Europe, this species may hunt during the day (Mikkola 1983).

Whereas the Long-eared Owl was long regarded as a vole *(Microtus)* specialist, more recent studies have shown that it is in fact very opportunistic. Although the diet is indeed mostly made up of voles and deer mice *(Peromyscus)*, a variety of other small mammals are taken, including heteromyid rodents such as pocket mice *(Perognathus* and *Chaetodipus)*, as well as shrews *(Blarina* and *Sorex)*, pocket gophers *(Thomomys)*, and even bats *(Antrozous* and *Myotis*, for example) and juvenile rabbits *(Lepus* and *Sylvilagus)*. One study site in Spain, a communal roost, yielded pellets that showed bats forming 2 percent of the diet of that particular population of Long-ears, which preyed on these mammals regardless of the availability of rodents, the main food (García et al. 2005). Birds, in particular small passerines, can form an important dietary component.

Occasionally, this owl eats reptiles as well as insects. One Long-eared Owl was observed hawking flying moths (Sleep and Barrett 2004). Still, given a variety of choices, Long-ears focus on a few species, as shown in studies at desert winter roosts and elsewhere (Barrows 1989; Brown 1995; Marti 1976). More than a dozen wintering Long-ears in Germany during a period of deep snow quickly learned to line up on garden fence posts to await the delivery of live laboratory mice provided on a tray by owl lovers.

REPRODUCTION: The male Long-eared Owl initiates courtship with display flights over potential nest sites. Zigzag flights through and around nest groves are interspersed with wing-clapping and advertising songs. The male may hover above the female as she inspects the nest, and the two vocalize back and forth.

This species, like most other owls, uses existing facilities—the old stick nests of magpies *(Pica)* (a site commonly used in the West), ravens and crows *(Corvus)*, various hawks, herons, squirrels, and even natural platforms in Arizona saguaros *(Cereus giganteus)* (Millsap 1998)—and it may nest in loose colonies and may share the same hunting grounds with nesting neighbors. In Orange and San Diego counties in California, old nests of Cooper's Hawks *(Accipiter cooperii)* and American Crows *(Corvus brachyrhynchos)* in closed-canopy live oak woodland were used by Long-ears, sometimes with the crows or hawks concurrently nesting nearby (Bloom 1994), associations also observed in San Benito County in moderately closed oak–gray pine woodland. The Long-eared Owl rarely nests on the ground or in tree or rock cavities (Marks et al. 1994).

Only the immediate vicinity of the nest is defended. Copulation takes place after the male, having put on an aerial and vocal performance, begins to sway while raising and lowering his wings. Apparently, the female finds this irresistible. DNA fingerprinting of 59 nestlings from 12 nests revealed that their mothers had not played around: opportunity notwithstanding, monogamy ruled (Marks, Dickinson, and Haydock 1999).

The Long-eared Owl lays its eggs from late February to mid-May. Although as many as 10 eggs may be produced, a typical clutch consists of four or five (or occasionally six) eggs, laid at two-day intervals. Incubation lasts about four weeks, and the young owls, still flightless, leave the nest at about three weeks of age, about one and a half to two weeks before they can take to the wing: when five weeks old, they are capable of short flights. The male Long-eared Owl feeds the young throughout and is joined by the female in this task when the owlets are about two weeks old. But she deserts them at the age of six and a half to eight weeks, when feeding once again is entirely up to the male for another two to three weeks. The young disperse from the nest area when they are about 10 to 11 weeks old.

DISTRIBUTION AND HABITAT: The Long-eared Owl breeds from eastern and southern British Columbia across the continent to Nova Scotia, south through most of California and the rest of the West, including south-central Arizona, northern New Mexico,

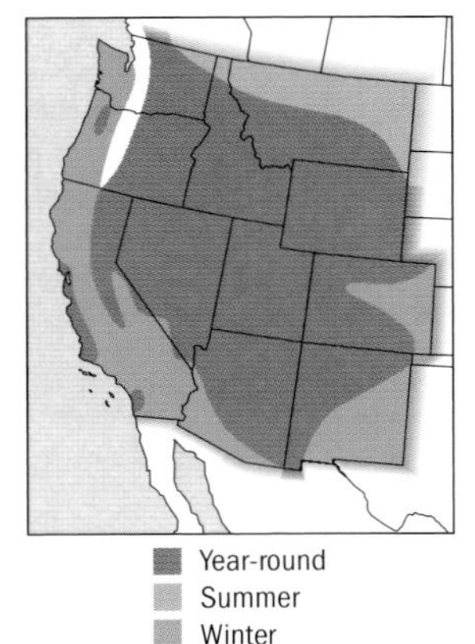

and northern Texas. Its range continues eastward through the central states to New Jersey and Appalachia.

In California, breeding owls of this species are scarce, with the exception of the riparian areas of the Susan River and other creeks near Honey Lake (Lassen County) and in the Owens Valley (Inyo County). Breeding occurs in the inner Coast Ranges and probably the outer ranges, too, of central California and in parts of the south coast as well as in the Cascades, the Modoc Plateau (where it may reach its current center of abundance [B. Woodbridge, pers. comm. 2006]), on the lower west slope of the Sierra Nevada, in the Great Basin area east of the Sierra, on isolated mountain ranges in the south and southeast of the state, and along the Mojave River. It is uncommon in the humid northwest.

It nests in a variety of habitats, importantly riparian woodland (including streams in Spotted Owl habitat), but also in dense live oaks *(Quercus)* next to grasslands, in blue oak–gray pine woodland, along the borders of the Great Basin sage steppe in cottonwoods *(Populus)*, willows *(Salix)*, and isolated junipers *(Juniperus)* (for example, in Lassen and Modoc counties), in clumps of tamarisk *(Tamarix)*, and even in dense coniferous forest of the Sierra Nevada near large meadows. In a southern California study, the majority of 69 historic nests were in oaks, with nearly as many in willows and in cottonwoods and much smaller numbers in a variety of trees such as eucalyptus *(Eucalyptus)*, sycamore *(Platanus)*, and walnut *(Juglans)*; 29 more recent nests were, with but one exception, in oaks (Bloom 1994), perhaps reflecting a shrinking population. The proximity of marshes or grasslands or low shrub steppe is a decisive factor.

For many years, a reliable place in California to see a Long-eared Owl has been at Tamarisk Campground in Anza-Borrego State Park (San Diego County), where the species not only winters but also nests, indifferent to people walking below the roost. Another reliable California winter roost site is at Mercey Hot

Fig. 122. One of several Long-eared Owls roosting in late winter in a tamarisk *(Tamarix)* grove in Anza-Borrego State Park (San Diego County, California), a reliable place to see this species.

Springs (Fresno County) from October to late spring. Communal roosts have been found in junipers in Lassen County.

The term "migration" does not seem to fit this species, for Long-ears may breed in one place one year and not return the next, sometimes nesting substantial distances from the previous year. Three adults, banded one nesting season, were recovered more than 450 km (280 mi) away during the next (Marks et al. 1994), and one that had been banded in Saskatchewan turned up in Oaxaca, 4,000 km (2,500 mi) away (Houston 1966). This owl may be more properly termed nomadic (P. Bloom, pers. comm. 2006) or irruptive.

The movements of Long-eared Owls in California are puzzling and often cover no great distance. They appear to wander downslope from the Sierra foothills in fall and congregate in nomadic flocks in winter, although some birds or populations may migrate (Zeiner et al. 1990). Unexpectedly, one individual banded in California in April was recovered in Ontario, Canada, in October (Marks et al. 1994). The species appears as a rare migrant on three of the Channel Islands, off the California coast.

SIMILAR SPECIES: Although the Western Screech-Owl *(Megascops* *kennicottii)* also has ear tufts that are normally conspicuous, it is far smaller than the Long-eared Owl, is dumpy looking, has much shorter tufts, and has a facial disk that is gray rather than rusty or buffy. Conversely, the Great Horned Owl *(Bubo virginianus)*, which can have a similarly colored facial disk, is far larger, its tufts are widely separated (as opposed to close together in the Long-ear), its belly is finely barred, not barred and streaked, and it lacks the Long-eared Owl's checkered pattern.

STATUS: Formerly common in California, breeding populations have been in decline for many decades for reasons that are unclear. It is listed as a California species of special concern but has no conservation designation in any other western state.

REMARKS: With its very long ear tufts and attractive plumage, the Long-eared Owl is an exotic-looking bird that because of its secretive nature often escapes notice. It is probably most commonly seen in winter, when sometimes dozens of birds roost together, often in isolated groves of trees where they are more readily noticed. In Duluth, Minnesota, this species is commonly seen in flight, migrating just after sunset, at an altitude of 30 to 50 m (100 to 150 ft) (Marks et al. 1994).

A female Cooper's Hawk *(Accipiter cooperii)* was discovered incubating two Long-eared Owl eggs and three of her own in an old hawk's nest, with the dispossessed owl unhappily perched nearby. Only hawks fledged from the nest (Bloom 1994).

Fig. 123. Defense-threat posture of the Long-eared Owl.

SHORT-EARED OWL

Asio flammeus

Pls. 17, 20

FIGS.: 39, 68, 124–128

IDENTIFICATION: A midsized, sometimes plump-looking owl, about as large as a crow. Its short ear tufts are very rarely obvious (usually when the owl feels aggressive) but mostly invisible or just barely visible. Long winged, it has a short tail and is overall the color of dry grass.

The head of the Short-eared Owl is ocher or buff with broad black brown streaks and chevrons. The facial disk is quite round, ocher with radiating thin black shaft streaks, with much black around the eyes like an exuberant application of mascara. Eyebrows and rictal fans are white to off-white, and there is a white triangle below the beak. Most or all of the disk is narrowly bordered with a rim of white feathers and, beyond, blackish ones.

The upperparts of this owl are dark brown mottled with ocher or buff, streaked in parts, often appearing as a bold pattern of

Fig. 124. Sibling Short-eared Owl young, demonstrating asynchronous hatching.

blackish triangles bordered by white or ocher. Underparts are buff to whitish (in males), with broad brown shaft streaks on the upper breast changing into progressively narrower streaks toward and on the belly. Tarsi and feet are densely feathered, the latter appearing small for a bird of this size.

The juvenile Short-eared Owl is dorsally dark brown, the feathers ending in buffy tips. The facial disk is black brown except for the white or mottled white and brown eyebrows, white rictal fans with fine black shaft streaks, and white chin. The underparts are ocher to buff, with narrow dark brown bars on the upper breast.

VOICE: A bark call, directed at humans that get too close, is probably the most often heard sound (Holt and Leasure 1993). It is a variant of the "keee-ow" call, which in winter may be addressed to other Short-eared Owls (Clark 1975) or at diurnal raptors. One advertising song is a series of more than a dozen "hoo-hoo-hoo" calls, delivered from the ground or a fence post or such, or in flight. During nest defense, both sexes may scream, bark, or whine (Holt and Leasure 1993). The young beg with hissing sounds.

FLIGHT: Two dark patches are prominent on the spread wing: one formed by the dark greater upper primary coverts (on the wing's upper surface) and the other on the lower primary coverts (on the underwing). Both patches are near the bases of the outer primaries, which themselves are conspicuously dark tipped.

Flying, this owl raises its wings with a little jerk, and then brings them down more slowly, making the flight look bouncy and at the same time buoyant. In foraging flight, it typically quarters close to the ground and sometimes hovers or kites in a wind. It also on occasion soars like a hawk and, when disturbed in its winter grounds, rises in arcs, sometimes to considerable heights, alternately flapping and gliding on slightly dihedral wings, to eventually descend in the same fashion. A spectacular courtship flight (see "Reproduction," below) puts it in a league with some diurnal raptors.

DAILY ACTIVITY PATTERN AND FEEDING: The Short-eared Owl can be active both day and night. It usually roosts (and sleeps) on the ground, concealed by tall grass or other vegetation. In winter, it often roosts communally (which likely promotes early detection of predators), and it may sometimes roost in trees (perhaps because of snow on the ground), occasionally in the company of Long-eared Owls *(A. otus)* (Bosakowski 1986). Other roost

Fig. 125. This Short-eared Owl has just caught a vole.

sites include quarries, gravel pits, and wrecked cars (Clark 1975).

Although the Short-eared Owl appears to be a diurnal hunter, this impression may simply result from its great visibility when it forages by day. It certainly does hunt during the day, but most likely when it has young to feed or when prey abundance is low. It is then often seen in late afternoon. In early May at Tule Lake in northern California, Short-eared Owls hunted chiefly between eight and 10 o'clock in the morning, probably to feed their hungry young (Dixon and Bond 1937). On the other hand, owls foraging in a salt marsh adjacent to a dump in Contra Costa County, California, did so almost entirely at night, perhaps to avoid mobbing by gulls drawn to the dump (Johnston 1956). Obviously, the time of day when the prey is active is also of prime importance. In winter, this owl is principally a crepuscular hunter (Clark 1975).

Most of the foraging, by far, is done on the wing, the owl flying at fairly low speed about 2 m (6 ft) above ground, sometimes hovering (as high as 30 m [90 ft]), and dropping onto prey. It may on occasion also hunt from a perch. Hearing probably plays the major role in locating prey: this owl's ear openings are asymmetrical in position, though similar in size.

The most commonly eaten food is voles *(Microtus)* in most of North America, although other small mammals, such as shrews *(Sorex* and *Blarina)*, pocket gophers *(Thomomys)*, pocket mice *(Perognathus* and *Chaetodipus)*, and a variety of other mice, kangaroo rats *(Dipodomys)*, and even rabbits *(Lepus* and *Sylvilagus)*, presumably juveniles, are also taken. In addition, the Short-eared Owl eats birds, especially in coastal areas; for example, on Bolinas

Fig. 126. A Short-eared Owl exhibits comfort behavior, simultaneously stretching both wing and leg on the left side.

Lagoon (Marin County), the contribution of birds to the diet of wintering Short-eared Owls was 87.9 percent in one study, chiefly Dunlins *(Calidris alpina)* (Page and Whitacre 1975). Although most avian prey consists of small passerines such as Savannah Sparrows *(Passerculus sandwichensis)* and American Pipits *(Anthus rubescens),* these owls fed on birds as large as adult Common Terns *(Sterna hirundo)* and Laughing Gulls *(Larus atricilla)* (Holt 1993). Insects, too, are eaten, including orthopterans and carabid beetles.

Small mammals are consumed whole or with heads removed, and small birds are swallowed minus their wings, whereas large ones are decapitated (sometimes also with extremities taken off) and the meatiest portions of the breast then eaten without prior plucking (Holt and Leasure 1993).

REPRODUCTION: Highly territorial, the Short-eared Owl aggressively defends its territory, which serves for both hunting and nesting, with prey abundance governing territory size (Lockie 1955).

The male Short-ear seeks a partner early in the year as the communal roosts break up, and his efforts to impress a female take the form of a spectacular sky-dance that may be performed at night and during the day. Rising as high as 140 m (450 ft) in small circles, the male then performs a shallow stoop ending in an upward swoop during which he claps his wings below his body, producing a sound that has been described as resembling "the flutter of a small flag in a very strong wind" (DuBois 1924). He also kites on spread wings and tail and sings his advertising song before he dives. Meanwhile, the female, perched below, calls and sometimes joins the male in his performance. Talon grappling has been observed, too (Hamerstrom et al. 1961). After several bouts of song and dive, the male descends with wings in a dihedral, rocking from side to side. One observer watched a male catch a vole and present it to the female with a song and a side-to-side rocking dance; she accepted the gift and presented herself for copulation.

The nest, built by the female on the ground, consists of a simple bowl scratched into the substrate and lined with grass and feathers. The site is typically a little elevated—on a small knoll, hummock, or such—and well vegetated with grasses and forbs or low perennials. Even stubble and hayfields may be used (Clark 1975), as are cattail *(Typha)* and cord grass *(Spartina)* marshes. Generally, the vegetation needs to be tall enough to conceal the incubating female.

The female lays five or six eggs (chiefly in April but as early as late March or as late as June in the Arctic) at one- to two-day intervals, although in years of vole population peaks, she usually produces many more. They are incubated, starting with the first laid, by her alone, for three and a half to a little over four weeks. The owlets develop unusually fast and very rapidly gain weight. They leave the nest at 14 to 17 days of age, in the sequence they were hatched, long before they can fly. These young pedestrians travel as much as 55 m (180 ft) from the nest. The rapid sequence of events reduces the chances of having a predator wipe out the entire brood, always a risk to a ground-nesting bird. The young begin to fly at about five weeks of age. They roost together, at first also with their parents, until they disperse.

DISTRIBUTION AND HABITAT: The Short-eared Owl is one of the most globally widespread of all owls. Partial to cool latitudes, it breeds in southern South America, a handful of islands around the world, across northern Eurasia and the northern half of our continent, throughout Alaska but chiefly below the Arctic Circle in Canada, to the southern border of their breeding distribution running roughly from southern California northeast to Virginia,

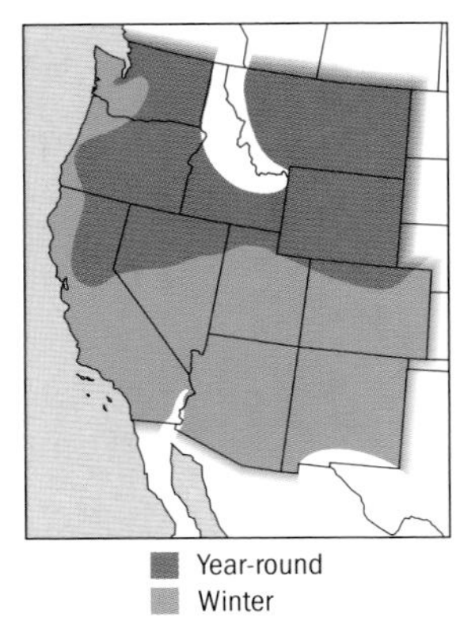

with the bulk of the population in the western states.

Although this species formerly occurred as a resident locally throughout California (except in higher mountainous country), today, Short-eared Owls breed in very few areas with any regularity. Some nest in the Klamath Basin (Siskiyou County), the Modoc Plateau, and some parts of the Great Basin and near Honey Lake (Lassen County) in the state's northwest. They have nested in the western Sierra Nevada below 600 m (2,000 ft) and in the San Francisco Bay Region and near Monterey. They are very rare nesters in the Central Valley and apparently no longer breed in the southern coastal areas but may do so in the North Mojave–Harper Dry Lake region. They are occasional breeders on some of the Channel Islands.

Short-eared Owls, however, are substantially more numerous in winter, when they migrate into California, and can be seen regularly, though not in as great numbers as formerly, in the Klamath Basin, in the Grizzly Island Wildlife Area, in the Sacramento Delta area (Solano County), and also in the Livermore Valley (Alameda County), where it is possible in some years to find dozens roosting together. Winter migrants may turn up elsewhere in California in suitable habitat. Migrants visiting the state arrive in September and October and leave in late March to early April.

Open areas with low vegetation are the domain of the Short-eared Owl. Annual or perennial grasslands, fresh- and saltwater marshes, irrigated pastures and prairies, and even alfalfa fields provide habitat for this owl, provided there is an abundance of voles.

Because it is nomadic, this species may unexpectedly appear in some numbers in rodent-rich areas, and migrating individuals appear in California deserts and have even been collected in Death Valley. Short-ears may travel great distances. An adult banded in British Columbia was found in California, 1,730 km (1,075 mi) to the southeast (Clark 1975).

SIMILAR SPECIES: The related Long-eared Owl *(A. otus)* is smaller but has much larger ear tufts that are usually prominent, and it has a checkered, not streaked belly. It also does not habitually sit on the ground. The similar-sized Barn Owl *(Tyto alba)* is also often found in open areas but is almost strictly nocturnal and has a heart-shaped (rather than round) facial disk, dark eyes, and virtually unmarked underwings. The Burrowing Owl *(Athene cunicularia)*, another open-country species, is much smaller, has conspicuously long legs, and has barred, not streaked, underparts.

In flight, the Short-eared Owl and the Long-eared Owl are difficult to tell apart. The Short-ear also might be mistaken for a Northern Harrier *(Circus cyaneus)*, a diurnal raptor that is found in the same habitat and forages in a similar manner. This hawk, however, is long tailed and has a very conspicuous white rump, and it is active only during the day.

STATUS: The Short-eared Owl is declining and is designated in California and Wyoming as state species of special concern, in Utah and Idaho as state species of concern, and in Nevada as a species of conservation priority. It is not listed federally.

REMARKS: The Short-eared Owl provides the owler with an unusual pleasure: not only is it often active in daylight, but, unlike

Fig. 127. A kettle of flying Short-eared Owls performing an aerial ballet.

its kin, this owl is highly aerial. It makes for a wondrous experience to see a dozen or more of these birds, 200 or so feet up in the cloud-dappled sky, engage in a stately airborne ballet that looks oddly formal as they glide about one another in arcs, intermittently beating their long, pale wings.

Large groups may roost together in winter in concealing grass as close as 1 m (3 ft) apart (Holt and Leasure 1993) but flush when approached by oblivious hikers or dogs, leaving oblong depressions or tunnels in the grass often resembling the forms made by jackrabbits *(Lepus)*. The large size of the form and the presence of owl whitewash and pellets, however, leave no doubt that Short-eared Owls were recently present. Multiple pellets and fecal accumulation suggest that these quarters are used for more than one day. One individual's form was completely hidden under long, flopped-over grass and held a pile of eight pellets on one end and a cake of many droppings at the other.

In Fresno County in 1913, an estimated 200 individuals were observed hunting together (Bent 1961), but winter aggregations so large are not likely to be seen today. In the Netherlands in 1948, 2,000 of the owls as well as hundreds of hawks were attracted to one area with a superabundance of rodent prey, but such numbers were not observed again (Voous 1989).

This species has also been called Marsh Owl or Prairie Owl.

Fig. 128. Short-eared owls rising from a grassy field in Alameda County, where they form communal winter roosts. At least six owls can be seen flying in this photo.

BOREAL OWL

Aegolius funereus

Pls. 18, 21

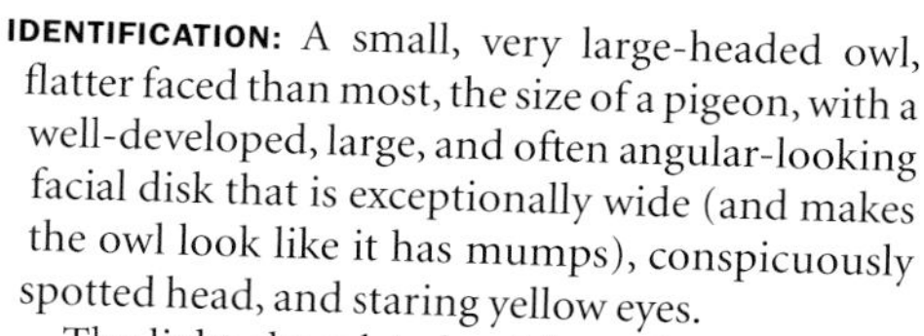

FIG.: 129

IDENTIFICATION: A small, very large-headed owl, flatter faced than most, the size of a pigeon, with a well-developed, large, and often angular-looking facial disk that is exceptionally wide (and makes the owl look like it has mumps), conspicuously spotted head, and staring yellow eyes.

The light chocolate head is profusely spotted with creamy white (in Sweden, this species is called the Pearl Owl). The white eyebrows and white areas on the gray and brown disk form a broad keyhole pattern that surrounds the eyes, beak, and black and brown rictal bristles. Often, the crown feathers above the eyes are raised to form corners somewhat like ear tufts (which are, however, absent); they impart a sometimes angular appearance to this owl's face that is also seen less pronouncedly in the Northern Saw-whet Owl *(A. acadicus)*. The beak is dull yellow to brownish.

The upperparts are light chocolate with prominent white spots, the largest running down the outer scapulars. The underparts are white with diffuse broad russet streaks that form a broken band across the upper breast. Both legs and toes are densely feathered.

The juvenile Boreal Owl is strikingly sooty black brown on the head and upperparts, with a black facial disk that sharply sets off the white to gray eyebrows and upper rictal bristles. The underparts are pale brown and heavily marked with diffuse dark brown streaks. Colored entirely differently from the adult, the young owl suggests a separate species.

VOICE: The advertising song is a series of five to eight liquid or staccato "hoops" (Voous 1989), heard most often in late winter or early spring, especially during a full moon (Dunn 1988). In Norway, an estimated 4,000 hoots represented one male's vocal effort for one night. Other calls include trills and barks (Holt et al. 1999), and the young's begging call is "psee."

FLIGHT: A forest dweller, this owl flies between perches in direct flight.

DAILY ACTIVITY PATTERN AND FEEDING: Mostly nocturnal in its foraging, the Boreal Owl spends the day roosting in dense foliage

Fig. 129. A juvenile Boreal Owl, like other owls, handles small food items with one foot, parrot style.

on steep slopes, often next to a tree trunk (Hayward and Garton 1984).

This owl moves in a zigzag pattern as it progresses through the forest still-hunting. It may wait 10 minutes or more before attacking, expecting the prey to make itself more vulnerable (Holt et al. 1999). It also plunges through shrubs to catch its quarry (Palmer 1986). A Boreal Owl moves prodigious distances in winter in search of scarce food, and, unlike the day roosts of other owls, which are often the same day after day, those of the Boreal may be miles apart, the owl almost never reusing the same tree on consecutive days (Hayward and Hayward 1989). At times, this species forages in more open areas, even clear-cut areas (in Europe), if that is where the prey is.

The Boreal Owl feeds chiefly on small mammals such as voles *(Clethrionomys* and *Microtus)* and shrews *(Sorex),* small birds, and insects (Johnsgard 2002). It can readily locate prey under snow and other cover, reflecting its well-developed auditory system.

REPRODUCTION: A Boreal Owl does not defend a territory, and home ranges can greatly overlap—unusual behavior for owls. By vocalizing, a male attracts a female to nest cavities, made usually by Northern Flickers *(Colaptes auratus)* and Pileated Woodpeckers *(Dryocopus pileatus),* but little is known about courtship.

Egg-laying takes place from about mid-April to early June. In the western United States, the female lays from two to four eggs (in Fennoscandia, five to nine eggs in good years [Hayward and Hayward 1989]) and incubates them for about four to four and a half weeks, the young prefledging at four to five weeks of age and reaching independence three to six weeks after leaving the nest (Holt et al. 1999). In Idaho, males foraging to feed the females and young hunted at higher elevations where prey was several times more common than lower down where the nests were located. Apparently, the dearth of nest sites higher up forced these hunting commutes (Hayward and Hayward 1989).

The female Boreal Owl is known to abandon her half-grown young and partner to mate again with another male, behavior known as serial polyandry.

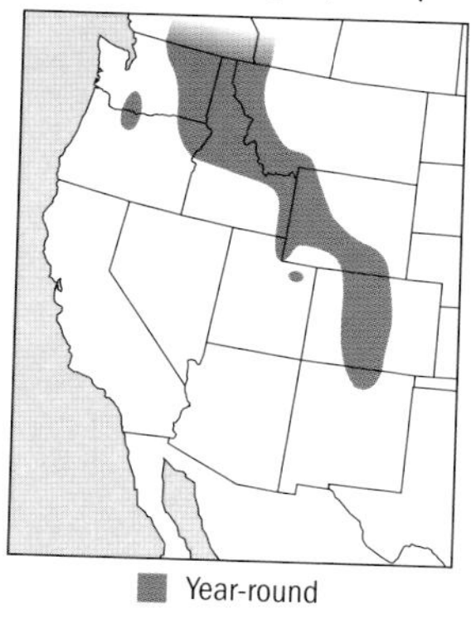

DISTRIBUTION AND HABITAT: The Boreal Owl is obviously a bird of boreal forests, where it is the most common owl species (Hayward and Hayward 1989), although it breeds southward in mountain ranges both in North America and in Eurasia.

On our continent, its chief breeding distribution follows the broad latitudinal band of coniferous forest that spans North America from central Alaska to the Canadian maritime. Southward breeding populations extend through western Alberta and northern British Columbia and occur spottily in Minnesota, Montana, Wyoming, Colorado, New Mexico, Idaho, Washington, and Oregon, including the Blue Mountains in the northeast of that state.

Boreal Owl experts believe it is probable that the species breeds in the mountains of northern California (Hayward and Hayward 1993), and indeed it is possible that it may be discovered the length of the Sierra Nevada. Northern populations, food-stressed in winter, move into southern Canada and the northern and northeastern states in numbers (Catling 1972). Such irruptions may on occasion penetrate northern California. Wintering birds have been found close to the California border, and one may have been heard calling at Echo Lake (El Dorado County) during a winter night in late 1985 (Dunn 1988).

Dense spruce-fir forests above 1,500 m (5,000 ft) are frequent habitat for these owls in the western states, and they were found in subalpine fir *(Abies lasiocarpa)* and in western hemlock *(Tsuga heterophylla)* forests in Idaho and Montana (Hayward et al. 1987). In a Colorado study, the owls were most numerous above 2,750 m (9,000 ft) in mature spruce-fir forests with many subalpine meadows (Palmer 1986).

In one Canadian study, the owls roosted in trees such as the balsam fir *(Abies balsamea)*, whose foliage formed a cocoon, the owls perched, quite exposed sometimes, on bare branches within (Bondrup-Nielsen 1978).

SIMILAR SPECIES: The Boreal Owl much resembles the Northern Saw-whet. The latter, however, is a lot smaller (robin-sized versus pigeon-sized), has distinct reddish brown streaks on its underparts, lacks the Boreal's bold black and brown frame around its facial disk, and has white streaks rather than spots on its crown. Other similar-sized forest owls have ear tufts.

STATUS: The Boreal Owl is a secretive species that lives in inaccessible areas, making it so difficult to find and to study so that its conservation status is largely unknown. The U.S. Forest Service considers it a sensitive species. It is listed as threatened by the state of New Mexico, is a species of special concern in Wyoming, and is labeled vulnerable in Idaho. It has not been officially recorded as a California species.

REMARKS: It is a measure of the Boreal Owl's elusiveness that it was not discovered nesting in the western United States until 1980, when a population was found in central Idaho (Hayward and Hayward 1989). Adding to the remoteness of its habitat are the owl's far-flung foraging excursions that make detection difficult. Finding breeding Boreal Owls can be problematic because they are not only nocturnal but also most vocal between February and April in their high-elevation snowbound haunts (Stahlecker and Duncan 1996).

In the Old World, this species is known as Tengmalm's Owl.

NORTHERN SAW-WHET OWL *Aegolius acadicus*

Pls. 19, 21

FIGS.: 8, 12, 24, 31, 37, 38, 59, 130–132

IDENTIFICATION: A very small yellow-eyed owl, about the size of a robin, with a very short tail and noticeably wide eartuftless head relative to body size, making it appear somewhat top heavy. The short legs are densely feathered, as are the toes. The female is slightly larger and about 25 percent heavier than the male.

The facial disk is symmetrical and conspicuous. Patterned like a sunburst, it bears buff, narrow rays on a brown background radiating out from the eyes that, when the latter are wide open, contribute to giving the bird a startled expression. White dashes demarcate the edges of the disk and decorate the owl's forehead between the disk's two lobes, as well as the crown and the nape. The eyebrows are white and join above the beak, and small patches of black feathers are conspicuous between the black beak and the eyes.

The upperparts are uniform tan, gray brown to brown except for two or three lines of white spots (with some partly hidden) running the length of the scapular tract, with additional spots on the upperwing coverts and along the leading edge of the wing. The white underparts bear broad reddish brown streaks that sometimes merge into a nearly solid broad breast band bordered above by a narrower white band or broad white triangle.

As in the Boreal Owl *(A. funereus)*, the juvenile is strikingly different from the adult, so much so that a tyro could easily consider it a separate species. The blackish facial disk emphasizes the stark white of the eyebrows conjoined above the beak. The remainder of the head and the upperparts are blackish brown with a slight reddish tinge. The brown breast contrasts with the unmarked yellowish buff belly and flanks. A fledgling just out of the nest was molting numerous juvenal contour feathers.

VOICE: The advertising song of the Northern Saw-whet is the most commonly heard and easily recognized vocalization of this species. It consists of a series of toots, about two per second (sometimes faster), on a constant pitch, like "the warning beep of a large truck backing up" (C. Lenihan, pers. comm. 2005), a helpful mnemonic because of the inclination to imitate this call at too low a pitch. It is both higher and faster than the toots of a Northern Pygmy-Owl *(Glaucidium gnoma)*, which, however, is also often mimicked at too low a pitch. When the Northern Saw-whet is excited, the final one or two toots are slightly drawn out into a quavering whinny. An agitated male uttered a protracted, high-pitched "meem" between a series of toots (perhaps the call described by others as a nasal whine).

The owl is capable of adjusting the calls' volume from very soft to very loud. By lowering the volume and turning its head, a Northern Saw-whet can "throw" its voice, making it appear to be coming from some distance away.

Alarm calls include single, growling barks and a rapid series of chirps very similar to those produced by an Audubon Bird Call. A fledged young gave two sharp barks, much like a terrier's, in response to the noise made by a low-flying single-engine plane. Another apparent alarm call was a sibilant "psit-psit." In addition, there are calls described as sharp squeaks (supposedly resembling the sounds produced by the sharpening of a large saw), hissing calls, buzzes and twitters, and even blood-curdling shrieks. A captive on occasion gave calls, of unknown function, that seemed identical to the song of a Common Poorwill *(Phalaenoptilus nuttallii)* and the flocking call of a Wild Turkey *(Meleagris gallopavo)*; the alarm notes of a House Finch *(Carpodacus mexicanus)*; and a drawn-out whine, "psee-psee," like the sound produced during the dive display of a male Costa's Hummingbird *(Calypte costae)*, though much higher in pitch. (See also "Young and Their Care" in the section "An Owl's Life.") The young beg with soft, almost whispering, high-pitched chirps.

Although the Northern Saw-whet can be heard throughout the year, it most actively sings in California in winter and early spring, from December through March. One owl called almost incessantly in January, day and night, for several days, driving a nearby resident almost to distraction (I. Tiessen, pers. comm. 2006).

FLIGHT: This owl has fast wingbeats and generally flies low to the ground, often in an undulating fashion like a woodpecker (Cannings 1993). The light wing-loading allows for agile flight in relatively dense cover. The wings are a bit more pointed than those of a screech-owl *(Megascops)*. In flight, as in the perched bird, the tail looks conspicuously short, indicating an inclination toward long-distance flying, perhaps reflecting the Northern Saw-whet's migratory, or at least nomadic, nature.

Fig. 130. By simply lowering its volume and turning its head as it sings, a Northern Saw-whet Owl can throw its voice, making it very difficult to find.

DAILY ACTIVITY PATTERN AND FEEDING: The Northern Saw-whet Owl has a talent for making itself invisible as it roosts during the day. It may insert itself into clumps of mistletoe *(Phoradendron)*, vine tangles, thickets, dense twiggery (especially in conifers), the thick lower edges of a tree's canopy (such as in live oaks *[Quercus]*), or perch next to a small tree trunk.

This owl forages almost exclusively at night, seeking out tree perches along forest edges or margins of clearings or, on migration, in more open areas, where it sometimes uses fence posts. It starts after usually ground-based prey by dropping from elevated perches. The light wing-loading of this species has led some authors to suggest that it is more maneuverable than related owls and can therefore hunt in shrubby terrain (Palmer 1986). The Northern Saw-whet feeds principally on small rodents, at least during the breeding season, and because of the habitat, it is a real deer mouse *(Peromyscus)* specialist, although, if the opportunity

Fig. 131. The enormous ear opening of a Northern Saw-whet Owl runs the height of the head. The curved, blue gray structure inside is the back of the right eye.

presents itself, it also feeds on voles *(Microtus* and *Clethrionomys)* and other mice, as well as on juveniles of larger rodents. Shrews *(Sorex, Blarina,* and *Cryptotis)* and other small mammals, together with insects, are also caught, along with a few small birds (and rarely large ones, chiefly taken on migration when they are active at night). In coastal British Columbia, these owls take numerous intertidal invertebrates, and Northern Saw-whets migrating in fall in Virginia ate a surprising number of insects, comprising 80 percent of the prey remains found in the stomachs of 15 dead owls (Whalen et al. 2000).

The very big ear openings and the strongly asymmetric development of the ears indicate that hearing plays an important role in foraging. A Northern Saw-whet commonly eats only the front half of a captured mouse and stores the remainder on a branch or holds it in one foot, at times for hours (Collins 1993).

REPRODUCTION: A male Northern Saw-whet Owl's courtship of a female recalls the romantic overtures of young humans. He circles her in the air a number of times, then, upon landing, produces an assortment of odd calls and goes on to perform an entertaining bobbing and shuffling dance with which he approaches the female. Sometimes he graciously deposits a dead mouse near her (Karalus and Eckert 1974). Although the male may sing from promising holes, it is the female that likely selects the nest site. Cavities made by Northern Flickers *(Colaptes auratus)* and Pileated Woodpeckers *(Dryocopus pileatus)* are commonly used and are most often in dead deciduous trees. These sites are typically 3 to 6 m (10 to 20 ft) above ground, although the owl readily avails itself of nest boxes mounted much lower on trees, including those put up for Wood Ducks *(Aix sponsa)* in close proximity to water.

The owl builds no nest, the female simply depositing from four to seven eggs (with five to six the most common) on the wood chips or other debris in the cavity. One successful nest in a box in Alameda County, California, consisted of four inches of moss, a few dead leaves, and a partly shredded mylar balloon, undoubtedly a wood rat's *(Neotoma)* bedding installed before the owls moved in. The female does all the incubating, for a period of about four weeks, during which the male supplies her with food, frequently to excess.

Egg-laying takes place between late February and mid-April, probably chiefly in March. Because the eggs are laid at intervals of up to three days and incubation commences with the first laid, the young hatch asynchronously. When the last-hatched chick is about 18 days old (and therefore well feathered), many females leave the nest to roost elsewhere or go forth to mate with another male to raise a second brood, leaving the task of raising the first set of young to husband number one. If such is the case, the older young may feed their younger siblings with the mice supplied by the male.

The young fledge after about four and a half to five weeks, doing so at intervals, with the youngest leaving the nest cavity last. The young of the Alameda County nest mentioned above fledged in mid-May. The siblings tend to remain together after fledging and are fed by the male (and sometimes the female) for at least a month (Cannings 1993).

DISTRIBUTION AND HABITAT: The breeding range of the Northern Saw-whet Owl extends from southern Alaska in a north-south band that widens from the coast through the mountains of the western states to the Mexican border, and in a second band from British Columbia eastward through the Canadian provinces, the north-central states, and the Great Lakes to the Atlantic Ocean. Smaller isolated breeding populations are found in Tennessee and North Carolina and in western Mexico. Although northernmost Northern Saw-whets withdraw from their breeding areas in winter, most populations remain on their home turf year-round, although the species is sufficiently migratory that it has turned up in all continental states.

In the mountains of the West, Northern Saw-whets breed chiefly in low- to midelevation forests of ponderosa pine *(Pinus*

Fig. 132. Expressions and attitudes of a juvenile Northern Saw-whet Owl.

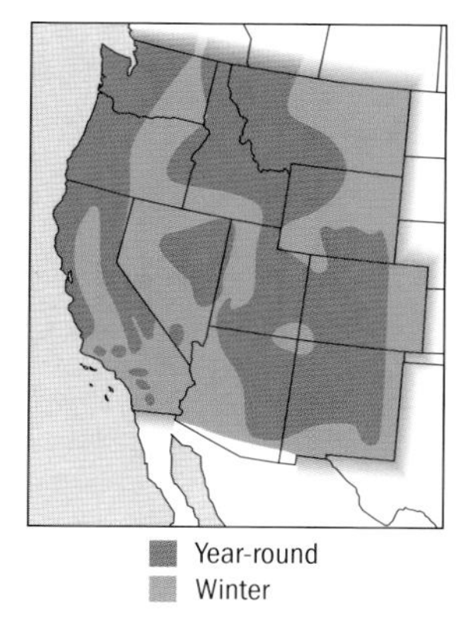

ponderosa), Douglas-fir *(Pseudotsuga menziesii)*, western redcedar *(Thuja plicata)*, fir *(Abies)*, and larch *(Larix)*, although they also nest in smaller numbers higher up. Surprisingly, they have also been found in nest boxes put up for other species in willow groves in the middle of the sage steppe of Idaho (J. Schmitt, pers. comm. 2007).

In California, the Northern Saw-whet Owl breeds wherever there are forests or woodlands, up to an elevation of about 2,450 m (8,000 ft), though it may wander even higher after breeding. In winter, it may be discovered nearly anywhere in the state (except the highest snowbound regions), including in parks, backyards, and golf courses, with some even turning up in desert oases and on the Farallon Islands west of San Francisco.

Breeding habitats in California include dense stands of unthinned Douglas-fir *(Pseudotsuga menziesii)* and tan-oak *(Litho-*

Fig. 133. Oak-bay-madrone forest with a fairly continuous canopy is excellent Northern Saw-whet habitat in central California.

carpus densiflorus) in Humboldt County (Hunter et al. 2005). In the Cascades, Klamath Mountains, and Sierra Nevada, the Northern Saw-whet breeds in pine-oak woodland, with ponderosa *(Pinus ponderosa)* and gray pine *(P. sabiana)* conspicuous species. Along the central coast, canyons with coast redwoods *(Sequoia sempervirens)* and oaks *(Quercus)* are popular. Although these habitats suggest a strong affinity to conifers, Northern Saw-whets in the Coast Ranges also nest in oak woodland or oak-bay woodland, perhaps preferring there is a fairly dense continuous canopy. In southern California, it lives in maul oak–black oak woodland mixed with conifers (P. Bloom, pers. comm. 2005), and nests have been found in some of the higher mountain ranges, such as the San Bernardino Mountains. In Kern County, it has even been found nesting in oak woodland–savannah with patches of chaparral (J. Schmitt, pers. comm. 2007). Certain Channel Islands also have breeding owls (albeit uncommon, sometimes in eucalypts), and riparian woodland furnishes breeding habitat in the Great Basin east of the Sierra Nevada and in the Sierra itself.

SIMILAR SPECIES: Lacking ear tufts and very distinctively marked, a Northern Saw-whet is unlikely to be mistaken for another species except perhaps for a Boreal Owl, which is larger, bigger headed still, and has a much more restricted range and habitat.

STATUS: As elsewhere in the West, the Northern Saw-whet Owl is a common to moderately common species in California in appropriate habitat and appears to be holding its own, logging operations in the northwest of the state and in the Sierras notwithstanding.

REMARKS: Because of its secretive habits, this owl is easily overlooked during the day. It tolerates extremely close approach and will allow a human to pass within a few feet without flying off (and thereby give away its presence); it can, in fact, sometimes be caught by hand. At night, it is famously eager to respond to a whistled imitation of its song and may do so even during the day from its hiding place, albeit with much less enthusiasm.

GLOSSARY

Adiposity The amount of fat tissue in the body.

Advertising song A vocalization, usually (but not always) given by a male, that proclaims the presence of a defended territory while simultaneously attracting a female. Also called *primary song.*

Allopreening Mutual preening between members of a mated pair or within a family.

Antiphonal A type of singing in which both pair members sing alternately, jointly producing a song.

Arboreal Living in trees.

Arthropod An invertebrate animal with a hard exoskeleton and jointed legs, for example, an insect, spider, or crab.

Barb A branch arising from the central shaft (rachis) of a feather.

Brood patch A thickened patch of skin denuded of feathers, on the belly of many birds, rich in blood vessels to supply heat during incubation and brooding.

Caching The hiding of killed or disabled prey for future use.

Carnivore Any meat-eating animal. A member of the mammalian order Carnivora is called a carnivoran.

Circumpolar Distributed in the regions encircling a pole.

Clutch The full set of eggs of a bird.

Colonial Living, and especially breeding, in the close proximity of conspecifics.

Commensalism A type of symbiotic relationship in which one organism associates with another at no expense to the partner. A relationship between organisms may be described as commensal.

Communal roost An assemblage usually of conspecifics, often gathered for the night.

Cone A specialized light receptor on the retina of the eye that perceives a color.

Conspecific An animal or plant belonging to the same species as the one under discussion.

Consumer pyramid A diagrammatic pyramid that illustrates the numbers of organisms feeding on each other in an ecosystem or a community.

Convergent evolution The evolution of similar traits by unrelated organisms having a similar lifestyle, for example, the development of talons in owls and hawks.

Copulation Mating.

Corvid A member of the crow family (Corvidae), such as a Steller's Jay *(Cyanocitta stelleri)*.

Differential hearing The perception of sound at a time interval between one ear and the other.

Dihedral The position of the wings, in flight, in which they are carried above the horizontal.

Displacement activity An activity unrelated to the situation at hand; for example, scratching one's head during an exam.

Diurnal Active during the day.

Ecosystem A major stable system produced by the interactions of organisms between themselves and their nonliving environment.

Exoskeleton A hard, armorlike covering of arthropods to which the muscles are attached from inside and that also serves as a water barrier.

Eyebrow The arc of an owl's facial disk that descends over the eye toward the beak.

Fledging Leaving the nest (by a young bird).

Form A depression or hollow made by jackrabbits *(Lepus)* and Short-eared Owls *(Asio flammeus)*.

Habitat The place where an organism lives; its address.

Heteromyid A rodent belonging to the distinctive family of kangaroo rats, kangaroo mice, and pocket mice. The literal meaning is *different mouse.*

Hovering Remaining stationary in the air by beating the wings, usually rapidly.

Hybrid An offspring resulting from the crossbreeding of two species, usually showing traits of both.

Immature A nonadult.

Infrared radiation The emission of light at a wave length invisible to the human eye, perceived as heat.

Insectivorous Insect eating.

Interspecific Describing behavior or interactions between different species, such as competition.

Intraspecific Describing behavior or interactions between members of the same species, such as competition.

Invertebrate An animal lacking a backbone or backbonelike structure.

Irruption year A year in which unusually large numbers of birds of a northern species invade more southerly latitudes.

Juvenal An adjective applying to the plumage or feathering of a bird in its first year (following one or more coats of down).

Kiting Remaining stationary in the air by the bird orienting its body into a strong wind, with little or no beating of the wings.

Mantling The covering of just-caught prey by a raptor with its wings, to hide it from potential robbers.

Mobbing Harassing of a predator (real or perceived) by one or more birds of one or more species, often with the aim of driving it away from a nest site or a nesting colony.

Morph An individual variant of a species, usually designated by color, for example, a rufous morph.

Mutualist An organism engaged in a symbiotic relationship with another, which results in benefits to both.

Nape The back of the neck.

Neotropics The tropical zone of the western hemisphere, also called *the New World tropics.*

Niche The entire ecologic relationship of an organism to its community.

Nictitating membrane The third eyelid, in reptiles and birds, which sweeps over the eye from the eye's inner corner, wiping and protecting it.

Nocturnal Active at night.

Operculum In biology, a lid or covering.

Orthopteran An insect belonging to the order of grasshoppers and crickets. The literal meaning is *straight-winged.*

Passerine A member of the bird order Passeriformes, meaning the *sparrow-shaped* birds, from kinglets to ravens.

Pellet An oblong, usually smooth packet of indigestible matter, such as fur, bones, insect parts, and accidentally ingested stems and leaves, regurgitated by hawks, owls, and some other birds.

Plumage All the feathers of a bird worn at a given time.

Population All the members of one species in a given area.

Preen To clean, oil, and order the feathers with the beak.

Prefledge To leave the nest before flying abilities are fully or sometimes even partly developed; common in owls.

Race A variant subpopulation of a species, adapted to a particular environment. If a race is well defined, it is equivalent to a subspecies.

Raptor Any bird of prey.

Rectrices The usually large tail feathers. The singular is *rectrix.*

Remiges The usually large flight feathers of the wing. The singular is *remex.*

Resident An animal that does not migrate; also, called *sedentary.*

Rictal bristle A small bristlelike feather of variable length that lacks barbs, found near the gape and cere, often covering the nares (nostrils).

Rictal fan The fanlike assemblage of bristles extending over the gape and beak.

Riparian Pertaining to the banks of bodies of freshwater.

Rod A light receptor on the eye's retina that is involved in black-and-white vision.

Roost A site where a bird or birds spend time, such as a day roost of owls.

Rufous A reddish brown or fox red color.

Salmonella A genus of bacteria, causative agent of an intestinal disease contracted from contaminated food or water. Rodents commonly are vectors.

Sclerotic ring A ring of small bone plates (ossicles) assembled in a collarlike ring to strengthen the eye and anchor its internal muscles in birds (and reptiles).

Secondary covert A small wing feather that overlies a secondary remex.

Sympatric Living in the same place or area with another organism, often said of two closely related organisms living side by side.

Thermoregulation The maintenance of a fairly constant body temperature by various means.

Triangulation The process of locating an individual or object by taking bearings from two or more different positions.

Trophic level A step in the movement of energy through an ecosystem, often illustrated by a pyramidal diagram.

Undulating Moving up and down, like waves in a row.

Vertebrate An animal having a backbone made up of a chain of connecting bones, the vertebrae.

Vole Any of several species of mouse-sized, usually short-tailed rodents, some of which are important raptor food.

Whitewash Raptor droppings, often conspicuous on cliffs because of the chalk white color of the uric acid contents.

Wing-loading The amount of weight that a unit area of wing must support. It is high in a Wild Turkey *(Meleagris gallopavo)*, and low in a Red-tailed Hawk *(Buteo jamaicensis)*.

REFERENCES

Allen, D.L. 1967. *The Life of Prairies and Plains.* New York: McGraw-Hill Book Co.

Allert, O.P. 1928. The last days of a certain Great Horned Owl. *Wilson Bulletin* 40:54.

Andersen, D.E. 1996. Intra-year reuse of Great Horned Owl nest sites by Barn Owls in east-central Colorado. *Journal of Raptor Research* 30 (2):90–92.

Appleby, B.M., and S.M. Redpath. 1997. Indicators of male quality in the hoots of Tawny Owls *(Strix aluco). Journal of Raptor Research* 31 (1):65–70.

Armstrong, E.A. 1975. *The Life and Lore of the Bird.* New York: Crown Publishers.

Askins, R.A. 2000. *Restoring North America's Birds: Lessons from Landscape Ecology.* New Haven: Yale University Press.

Ault, S.J., and E.W. House. 1987. Electroretinographic responses of the Great Horned Owl *(Bubo virginianus). Journal of Raptor Research* 21 (4):147–152.

Austen, M.J.W., M.D. Cadman, and R.D. James. 1994. *Ontario Birds at Risk: Status and Conservation Needs.* Port Rowan, ON: Federation of Ontario Naturalists and Long Point Bird Observatory.

Balda, R.M. 1993. Metabolic rate and evaporative water loss of Mexican Spotted and Great Horned Owls. *Wilson Bulletin* 105 (4):645–656.

Balda, R.P., B.C. McKnight, and C.D. Johnson. 1975. Flammulated Owl migration in the southwestern United States. *Wilson Bulletin* 87 (4):520–533.

Barrowclough, G.F., and R.J. Gutiérrez. 1990. Genetic variation and differentiation in the Spotted Owl *(Strix occidentalis). Auk* 107 (4):737–744.

Barrows, C.W. 1981. Roost selection by Spotted Owls: An adaptation to heat stress. *Condor* 83 (4):302–309.

Barrows, C.W. 1989. Diets of five species of desert owls. *Western Birds* 20:1–10.

Barrows, C., and K. Barrows. 1978. Roost characteristics and behavioral thermoregulation in the Spotted Owl. *Western Birds* 9:1–8.

Barton, N.W., and D.C. Houston. 1996. Factors influencing the size of some internal organs in raptors. *Journal of Raptor Research* 30 (4):219–223.

Beck, T.W., and R.A. Smith. 1987. Nesting chronology of the Great Gray Owl at an artificial nest site in the Sierra Nevada. *Journal of Raptor Research* 21 (3):116–118.

Beck, T.W., and J. Winter. 2000. *Survey Protocol for the Great Gray Owl in the Sierra Nevada of California.* Vallejo, CA: U.S. Department of Agriculture Forest Service, Pacific Southwest Region.

Belthoff, J.R., and A.M. Dufty Jr. 1997. Corticosterone and dispersal in Western Screech-Owls *(Otus kennicottii).* In *Biology and Conservation of Owls in the Northern Hemisphere: 2nd International Symposium* (February 5–9, Winnipeg, MB), General Technical Report NC190, eds. J.R. Duncan, D.H. Johnson, and T.H. Nicholls, 62–67. St. Paul, MN: U.S. Department of Agriculture Forest Service, North Central Research Station.

Bennett, J.R., and P.H. Bloom. 2005. Home range and habitat use by Great Horned Owls *(Bubo virginianus)* in southern California. *Journal of Raptor Research* 39 (2):119–126.

Bent, A.C. 1961. *Life Histories of North American Birds of Prey,* part 2. New York: Dover Publications.

Bibby, B. 2005. *Deeper Than Gold.* Berkeley, CA: Heyday Books.

Bloom, P.H. 1994. The biology and current status of the Long-eared Owl in coastal southern California. *Bulletin of the Southern California Academy of Science* 93 (1):1–12.

Bloom, P.H., J.L. Henckel, E.H. Henckel, J.K. Schmutz, B. Woodbridge, J.R. Bryan, R.L. Anderson, P.J. Detrich, T.L. Maechtle, J.O. McKinley, M.D. McCrary, K. Titus, and P.F. Schempf. 1992. The *dho-gaza* with Great Horned Owl lure: An analysis of its effectiveness in capturing raptors. *Journal of Raptor Research* 26 (3):167–178.

Blus, L.J. 1996. Effects of pesticides on owls in North America. *Journal of Raptor Research* 30 (4):198–206.

Boal, C.W., and B.D. Bibles. 2001. Responsiveness of Elf Owls to conspecific and Great Horned Owl calls. *Journal of Field Ornithology* 72 (1):66–71.

Boal, C.W., B.D. Bibles, and R.W. Mannan. 1997. Nest defense and mobbing behavior of Elf Owls. *Journal of Raptor Research* 31 (3): 286–287.

Boal, C.W., D.E. Andersen, and P.L. Kennedy. 2005. Productivity and mortality of Northern Goshawks in Minnesota. *Journal of Raptor Research* 39 (3):222–228.

Bondrup-Nielsen, S. 1978. Vocalizations, nesting and habitat preferences of the Boreal Owl *(Aegolius funereus)* in North America. M.S. thesis, University of Toronto.

Bonnot, P. 1928. An outlaw Barn Owl. *Condor* 30:320.

Bosakowski, T. 1986. Short-eared Owl roosting strategies. *American Birds* 40:237–240.

Bosakowski, T., and D.G. Smith. 1997. Distribution and species richness of a forest raptor community in relation to urbanization. *Journal of Raptor Research* 31 (1):26–33.

Botelho, E.S., and P.C. Arrowood. 1996. Nesting success of Western Burrowing Owls in natural and human-altered environments. In *Raptors in Human Landscapes,* ed. D.M. Bird, D.E. Varland, and J.J. Negro, 61–68. San Diego: Academic Press Limited.

Bowmaker, J.K., and G.R. Martin. 1978. Visual pigments and colour vision in a nocturnal bird *Strix aluco* (Tawny Owl). *Vision Research* 18:1125–1130.

Brooks, A. 1929. Pellets of hawks and owls. *Condor* 31 (5):222–223.

Brown, N. 1995. Notes on the winter roost and diet of Long-eared Owls in the Sonoran Desert. *Journal of Raptor Research* 29 (4): 277–279.

Bruce, M.D. 1999. Family Tytonidae (Barn-owls). In *Handbook of the Birds of the World,* vol. 5, ed. J. del Hoyo, A. Elliott, and J. Sargatal, 34–75. Barcelona: Lynx Edicions.

Buchanan, J.B., T.L. Fleming, and L.L. Irwin. 2004. A comparison of Barred and Spotted Owl nest-site characteristics in the eastern Cascade Mountains, Washington. *Journal of Raptor Research* 38 (3):231–237.

Bull, E.L., and J.R. Duncan. 1993. Great Gray Owl *(Strix nebulosa).* In *The Birds of North America,* no. 41, eds. A. Poole and F. Gill. Philadelphia: The Academy of Natural Sciences and The American Ornithologists' Union.

Bull, E.L., and M.G. Henjum. 1990. Ecology of the Great Gray Owl. General Technical Report PNW-GTR-265. Portland, OR: U.S. Department of Agriculture, Forest Service.

Bull, E.L., J.E. Hohmann, and M.G. Henjum. 1987. Northern Pygmy-Owl nests in northeastern Oregon. *Journal of Raptor Research* 21 (2):77–78.

Bunn, D.S., A.B. Warburton, and R.D.S. Wilson. 1982. *The Barn Owl.* Vermillion, SD: Buteo Books.

Burkholder, G., and D.G. Smith. 1988. Great Horned Owls *(Bubo virginianus)* nesting in a Great Blue Heron *(Ardea herodias)* heronry. *Journal of Raptor Research* 22 (2):62.

Cade, T.J. 1952. A Hawk Owl bathing with snow. *Condor* 54 (6):360.

California Department of Fish and Game. 2003. Bird species of special concern. Sacramento, CA: California Department of Fish and Game, Habitat Conservation Planning Branch. www.dfg.ca.gov/hcpb/species/ssc/sscbird/sscbird.shtml.

Cannings, R.J. 1993. Northern Saw-whet Owl *(Aegolius acadicus)*. In *The Birds of North America*, no. 42, eds. A. Poole and F. Gill. Philadelphia: The Academy of Natural Sciences and The American Ornithologists' Union.

Cannings, R.J., and T. Angell. 2001. Western Screech-Owl *(Otus kennicottii)*. In *The Birds of North America*, no. 597, eds. A. Poole and F. Gill. Philadelphia: Birds of North America, Inc.

Catling, P.M. 1972. A study of the Boreal Owl in southern Ontario with particular reference to the irruption of 1968/69. *Canadian Field-Naturalist* 86:223–232.

Cavanagh, P.M. 1990. Above-ground nesting by Burrowing Owls. *Journal of Raptor Research* 24 (3):68–69.

Chandler, C.R., and R.K. Rose. 1988. Comparative analysis of the effects of visual and auditory stimuli on avian mobbing behavior. *Journal of Field Ornithology* 59 (3):269–277.

Clark, R.J. 1975. A field study of the Short-eared Owl (*Asio flammeus* Pontoppidan) in North America. *Wildlife Monographs* 47: 1–67.

Collier, M.E.T., and S.B. Thalman, eds. 1996. *Interviews with Tom Smith and Maria Copa.* San Rafael, CA: Miwok Archaeological Preserve of Marin.

Collins, C.T. 1963. Notes of the feeding behavior, metabolism, and weight of the Saw-whet Owl. *Condor* 65:528–530.

Collins, C.T. 1993. A threat display of the Northern Saw-whet Owl *(Aegolius acadicus)*. *Journal of Raptor Research* 27 (2):113–116.

Cope, J.B., and J.C. Barber. 1978. Caching behavior of screech owls in Indiana. *Wilson Bulletin* 90 (3):450.

Coulombe, H.N. 1971. Behavior and population ecology of the Burrowing Owl *(Speotyto cunicularia)* in the Imperial Valley of California. *Condor* 73 (2):162–176.

Cramp, S., ed. 1985. *Handbook of the Birds of Europe, the Middle East, and North Africa*, vol. 4. Oxford, England: Oxford University Press.

Cromrich, L.A., D.W. Holt, and S.M. Leasure. 2002. Trophic niche of North American Great Horned Owls. *Journal of Raptor Research* 36 (1):58–65.

Dark, S.J., R.J. Gutiérrez, and G.I. Gould. 1998. The Barred Owl *(Strix varia)* invasion in California. *Auk* 115 (1):50–56.

de la Torre, J. 1990. *Owls: Their Life and Behavior.* New York: Crown Publishers.

Delaney, D.K., T.G. Grubb, and D.K. Garcelon. 1998. An infrared video camera system for monitoring diurnal and nocturnal raptors. *Journal of Raptor Research* 32 (4):290–296.

Dement'ev, G.P., and N.A. Gladkov, eds. 1966. *Birds of the Soviet Union,* vol. 1. Jerusalem: Israel Program for Scientific Translations.

Deppe, C., D. Holt, J. Tewksbury, L. Broberg, J. Petersen, and K. Wood. 2003. Effect of Northern Pygmy-Owl *(Glaucidium gnoma)* eyespots on avian mobbing. *Auk* 120 (3):765–771.

DeSante, D. F, and T.L. George. 1994. Population trends in the landbirds of western North America. *Studies in Avian Biology* 15:173–190.

DeSante, D.F., and E.D. Ruhlen. 1995. *A Census of Burrowing Owls in California, 1991–1993.* Point Reyes Station, CA: Institute for Bird Populations.

DeSante, D.F., E.D. Ruhlen, S.L. Adamany, K.M. Burton, and S. Amin. 1997. A census of Burrowing Owls in central California in 1991. In *The Burrowing Owl, Its Biology and Management,* Raptor Research Report no. 9, eds. J.L. Lincer and K. Steenhoff, 38–48. Lawence, KS: Allen Press.

DeSante, D.F., E.D. Ruhlen, and D.K. Rosenberg. 2004. Density and abundance of Burrowing Owls in the agricultural matrix of the Imperial Valley, California. *Studies in Avian Biology* 27:116–119.

Dixon, J.S., and R.M. Bond. 1937. Raptorial birds in the cliff areas of Lava Beds National Monument, California. *Condor* 39:97–102.

Drost, C.A., and R.C. McCluskey. 1992. Extirpation of alternative prey during a small rodent crash. *Oecologia* 92:301–304.

DuBois, A.D. 1924. A nuptial song-flight of the Short-eared Owl. *Auk* 41:260–263.

Duke, G.E., S. Jackson, and O.A. Evanson. 1993. Great Horned Owls do not egest pellets prematurely when presented with a new meal. *Journal of Raptor Research* 27 (1):39–41.

Duncan, J.R. 2003. *Owls of the World: Their Lives, Behavior, and Survival.* Buffalo, NY: Firefly Books.

Duncan, J. 2005. Breeding Long-eared Owls invade southern Manitoba in 2005. *Wingspan* 14(2):5–6.

Duncan, J.R., and P.A. Duncan. 1998. Northern Hawk Owl *(Surnia ulula)*. In *The Birds of North America,* no. 356, eds. A. Poole and F. Gill. Philadelphia: Birds of North America, Inc.

Duncan, J.R., and P.A. Lane. 1988. Great Horned Owl observed "hawking" insects. *Journal of Raptor Research* 22 (3):93.

Duncan, P.A., and W.C. Harris. 1997. Northern Hawk Owls *(Surnia ulula caparoch)* and forest management in North America: A review. *Journal of Raptor Research* 31:187–190.

Dunn, J.L. 1988. Tenth report of the California bird records committee. *Western Birds* 19 (4):129–163.

Duxbury, J.M., G.L. Holroyd, and K. Muehlenbachs. 2003. Large-scale dispersal of Burrowing Owls *(Athene cunicularia hypugaea)* as determined by stable isotope analysis. Paper presented at the 2003 Raptor Research Foundation Annual Meeting, Anchorage, Alaska.

Earhart, C.M., and N.K. Johnson. 1970. Size dimorphism and food habits of North American owls. *Condor* 72 (3):251–264.

Ehrlich, P.R., D.S. Dobkin, and D. Wheye. 1992. *Birds in Jeopardy: The Imperiled and Extinct Birds of the United States and Canada, Including Hawaii and Puerto Rico.* Stanford, CA: Stanford University Press.

Erickson, J.G. 1955. Pintails harassing a Short Eared Owl. *Auk* 72 (4):431.

Evens, J., and R. LeValley. 1982. Middle Coast Region. *American Birds* 36:890.

Everett, M. 1977. *A Natural History of Owls.* London: Hamlyn Publishing Group.

Fetz, T.W., S.W. Janes, and H. Lauchstedt. 2003. Habitat characteristics of Great Gray Owl sites in the Siskiyou Mountains of southwestern Oregon. *Journal of Raptor Research* 37 (4):315–322.

Feusier, S. 1989. Distribution and behavior of Western Screech-Owls *(Otus kennicottii)* of the Starr Ranch Audubon Sanctuary, Orange Co., California. M.S. thesis. Arcata, CA: Humboldt State University.

Flieg, G.M. 1971. Tytonidae × Strigidae cross produces fertile eggs. *Auk* 88 (1):178.

Folliard, L.B., K.P. Reese, and L.V. Diller. 2000. Landscape characteristics of Northern Spotted Owl nest sites in managed forests of northwestern California. *Journal of Raptor Research* 34 (2):75–84.

Forsman, E.D., E.C. Meslow, and H.M. Wight. 1984. Distribution and biology of the Spotted Owl in Oregon. *Wildlife Monographs* 87:1–64.

Forsman, E.D., R.G. Anthony, E.C. Meslow, and C.J. Zabel. 2004. Diets and foraging behavior of Northern Spotted Owls in Oregon. *Journal of Raptor Research* 38 (3):214–230.

Furniss, R.L., and V.M. Carolin. 1977. *Western Forest Insects.* Miscellaneous Publication 1339. Portland, OR: U.S. Department of Agriculture Forest Service, Pacific Northwest Range and Experimental Station.

Gaines, D. 1977. *Birds of the Yosemite Sierra: A Distribution Survey.* Oakland, CA: California Syllabus.

Gamel, C.M., and T. Brush. 2001. Habitat use, population density, and home range of Elf Owls *(Micrathene whitneyi)* at Santa Ana National Wildlife Refuge, Texas. *Journal of Raptor Research* 35 (3):214–220.

Ganey, J.L., and W.M. Block. 2005. *Dietary Overlap between Sympatric Mexican Spotted and Great Horned Owls in Arizona.* Research Paper RMRS-RP-57WWW. Fort Collins, CO: U.S. Department of Agriculture Forest Service, Rocky Mountain Research Station.

Ganey, J.L., W.M. Block, and R.M. King. 2000. Roost sites of radio-marked Mexican Spotted Owls in Arizona and New Mexico: Sources of variability and descriptive characteristics. *Journal of Raptor Research* 34 (4):270–278.

García, A.M., F. Cervera, and A. Rodríguez. 2005. Bat predation by Long-eared Owls in Mediterranean and temperate regions of southern Europe. *Journal of Raptor Research* 39 (4):445–453.

Gard, N.W., D.M. Bird, R. Densmore, and M. Hamel. 1989. Responses of breeding American Kestrels to live and mounted Great Horned Owls. *Journal of Raptor Research* 23 (3):99–102.

Garrett, K., and J. Dunn. 1981. *Birds of Southern California: Status and Distribution.* Los Angeles: Los Angeles Audubon Society.

Gehlbach, F.R. 1994a. *The Eastern Screech Owl: Life History, Ecology, and Behavior in the Suburbs and Countryside.* College Station, TX: Texas A and M University Press.

Gehlbach, F.R. 1994b. Nest-box versus natural-cavity nests of the Eastern Screech-Owl: An exploratory study. *Journal of Raptor Research* 28 (3):154–157.

Gehlbach, F.R. 1995. Eastern Screech-Owl *(Otus asio).* In *The Birds of North America,* no. 165, eds. A. Poole and F. Gill. Philadelphia: The

Academy of Natural Sciences and The American Ornithologists' Union.
Gehlbach, F.R. 1996. Eastern Screech Owls in suburbia: A model of raptor urbanization. In *Raptors in Human Landscapes,* eds. D.M. Bird, D.E. Varland, and J.J. Negro, 69–74. San Diego, CA: Academic Press Limited.
Gehlbach, F.R., and N.Y. Gehlbach. 2000. Whiskered Screech-Owl *(Otus trichopsis).* In *The Birds of North America,* no. 507, eds. A. Poole and F. Gill. Philadelphia: Birds of North America, Inc.
Gehlbach, F.R., and J.S. Leverett. 1995. Mobbing of Eastern Screech-owls: Predatory cues, risk to mobbers, and degree of threat. *Condor* 97 (3):831–834.
Giese, A.R., and E.D. Forsman. 2003. Breeding season habitat use and ecology of male Northern Pygmy-Owls. *Journal of Raptor Research* 37 (2):117–124.
Gifford, E.W., and G.H. Block. 1930. *California Indian Nights Entertainment.* Glendale, CA: A.H. Clark Company.
Gilliard, E.T. 1958. *Living Birds of the World.* Garden City, NY: Doubleday and Co.
Glutz von Blotzheim, U.N., and K.M. Bauer, eds. 1980. *Handbuch der Vögel Mitteleuropas,* vol. 9. Wiesbaden: Akademische Verlagsgesellschaft.
Goad, M.S., and R.W. Mannan. 1987. Nest site selection by Elf Owls in Saguaro National Monument, Arizona. *Condor* 89:659–662.
Grant, R.A. 1965. The Burrowing Owl in Minnesota. *Loon* 37:2–17.
Green, G.A. 1983. Ecology of breeding Burrowing Owls in the Columbia basin, Oregon. M.S. thesis. Corvallis, OR: Oregon State University.
Grimm, R.J., and W.M. Whitehouse. 1963. Pellet formation in a Great Horned Owl: A roentgenographic study. *Auk* 80 (3):301–306.
Grinnell, J. 1900. Birds of the Kotzebue Sound region, Alaska. *Pacific Coast Avifauna,* no. 1. Cited in Bent 1961.
Grinnell, J., and A.H. Miller 1944. *The Distribution of the Birds of California.* Berkeley, CA: Cooper Ornithological Club.
Gutfreund, Y., W. Zheng, and E.I. Knudsen. 2002. Gated visual input to the central auditory system. *Science* 297 (5586): 1556–1559.
Gutiérrez, R.J. 1996. Biology and distribution of the Northern Spotted Owl. *Studies in Avian Biology* 17:2–5.
Gutiérrez, R.J., A.B. Franklin, and W.S. Lahaye. 1995. Spotted Owl *(Strix occidentalis).* In *The Birds of North America,* no. 179, eds. A.

Poole and F. Gill. Philadelphia: The Academy of Natural Sciences and The American Ornithologists' Union.

Gutiérrez, R. J., J. E. Hunter, G. Chávez-León, and J. Price. 1998. Characteristics of Spotted Owl habitat in landscapes disturbed by timber harvest in northwestern California. *Journal of Raptor Research* 32 (2):104–110.

Guynup, S., and N. Ruggia. 2004. Owls face spotted future. *National Geographic News.* http://news.nationalgeographic.com/news/2004/07/0722_040722_tvspottedowl.html.

Haessly, A., J. Sirosh, and R. Miikkulainen. 1995. A model of visually guided plasticity of the auditory spatial map in the Barn Owl. In *Proceedings of the 17th Annual Conference of the Cognitive Science Society* (COGSCI-95, Pittsburgh, PA), 154–158. Hillsdale, NJ: Erlbaum.

Hamer, T. E. 1993. Hybridization between Barred and Spotted Owls. Abstract. *Journal of Raptor Research* 27 (1):72.

Hamer, T. E., D. L. Hays, C. M. Senger, and E. D. Forsman. 2001. Diets of Northern Barred Owls and Northern Spotted Owls in an area of sympatry. *Journal of Raptor Research* 35 (3):221–227.

Hamerstrom, F., F. Hamerstrom, and D. D. Berger. 1961. Nesting of Short-eared Owls in Wisconsin. *Passenger Pigeon* 23:46–48.

Hanna, W. C. 1954. Breeding dates for Barn Owls in southern California. *Auk* 71:90.

Hannah, K. C., and J. S. Hoyt. 2004. Northern Hawk Owls and recent burns: Does burn age matter? *Condor* 106 (2):420–423.

Hanson, R. 2000. Loving birds to death. *Audubon Magazine* 102 (May–June):18.

Haug, E. A. 1985. Observations on the breeding ecology of Burrowing Owls in Saskatchewan. M.S. thesis. Saskatoon: University of Saskatchewan.

Haug, E. A., B. A. Millsap, and M. S. Martell. 1993. Burrowing Owl *(Speotyto cunicularia).* In *The Birds of North America,* no. 61, eds. A. Poole and F. Gill. Philadelphia: The Academy of Natural Sciences and The American Ornithologists' Union.

Hayward, G. D., and E. O. Garton. 1984. Roost habitat selection by three small forest owls. *Wilson Bulletin* 96 (4):690–692.

Hayward, G. D., and E. O. Garton. 1988. Resource partitioning among forest owls in the River of No Return Wilderness, Idaho. *Oecologia* 75:253–265.

Hayward, G. C., and P. H. Hayward. 1993. Boreal Owl *(Aegolius funereus).* In *The Birds of North America,* no. 63, eds. A. Poole and F.

Gill. Philadelphia: The Academy of Natural Sciences and The American Ornithologists' Union.

Hayward, G.D., P.H. Hayward, and E.O. Garton. 1987. Movements and home range use by Boreal Owls in central Idaho. In *Biology and Conservation of Northern Forest Owls: Symposium Proceedings,* General Technical Report RM-142, eds. R.W. Nero, R.J. Clark, R.J. Knapton, and R.H. Hamre, 175–184. Washington, DC: U.S. Department of Agriculture Forest Service.

Hayward, P.H., and G.D. Hayward. 1989. Lone Ranger of the Rockies. *Natural History* 1989 (November):79–84.

Heinrich, B. 1987. *One Man's Owl.* Princeton, NJ: Princeton University Press.

Heizer, R.F., and A.B. Elsasser. 1980. *The Natural World of the California Indians.* Berkeley and Los Angeles: University of California Press.

Henry, S.G., and F.R. Gehlbach. 1999. Elf Owl *(Micrathene whitneyi).* In *The Birds of North America,* no. 412, eds. A. Poole and F. Gill. Philadelphia: Birds of North America, Inc.

Hensely, M.M. 1954. Ecological relations of the breeding bird population of the desert biome of Arizona. *Ecological Monographs* 24:185–207.

Herter, D.R., and L.L. Hicks. 2000. Barred Owl and Spotted Owl populations and habitat in the central Cascade Range of Washington. *Journal of Raptor Research* 34 (4):279–286.

Herting, B.L., and J.R. Belthoff. 1997. Testosterone, aggression, and territoriality in male Western Screech-Owls *(Otus kennicottii):* Results from preliminary experiments. In *Biology and Conservation of Owls in the Northern Hemisphere: 2nd International Symposium* (February 5–9, Winnipeg, MB), General Technical Report NC190, eds. J.R. Duncan, D.H. Johnson, and T.H. Nicholls, 213–217. St. Paul, MN: U.S. Department of Agriculture Forest Service, North Central Research Station.

Hoetker, G.M., and K.W. Gobalet. 1999. Predation on Mexican Free-tailed Bats by Burrowing Owls in California. *Journal of Raptor Research* 33 (4):333–335.

Holroyd, G.L., R. Rodríquez-Estrella, and S.R. Sheffield. 2001. Conservation of the Burrowing Owl in western North America: Issues, challenges, and recommendations. *Journal of Raptor Research* 35 (4):399–407.

Holscher, C.E. 1943. Observations on the Montana Horned Owl. *Condor* 45 (2):58–60.

Holt, D.W. 1993. Breeding season diet of Short-eared Owls from Massachusetts. *Wilson Bulletin* 105:490–496.

Holt, D.W., and S.M. Leasure. 1993. Short-eared Owl *(Asio flammeus).* In *The Birds of North America,* no. 62, eds. A. Poole and F. Gill. Philadelphia: The Academy of Natural Sciences and The American Ornithologists' Union.

Holt, D.W., and J.L. Petersen. 2000. Northern Pygmy-Owl *(Glaucidium gnoma).* In *The Birds of North America,* no. 494, eds. A. Poole and F. Gill. Philadelphia: Birds of North America, Inc.

Holt, D.W., L.J. Lyon, and R. Hale. 1987. Techniques for differentiating pellets of Short-eared Owls and Northern Harriers. *Condor* 89:929–931.

Holt, D.W., R. Kline, and L. Sullivan-Holt. 1990. A description of "tufts" and concealing posture in Northern Pygmy-Owls. *Journal of Raptor Research* 24 (3):59–63.

Holt, D.W., R. Berkley, C. Deppe, P.L. Enríquez Rocha, J.L. Petersen, J.L. Rangel Salazar, K.P. Segars, and K.L. Wood. 1999. Family Strigidae species accounts. In *Handbook of the Birds of the World,* vol. 5, eds. J. del Hoyo, A. Elliott, and J. Sargatal, 151–229. Barcelona: Lynx Edicions.

Hörnfeldt, B., T. Hipkiss, and U. Eklund. 2005. Fading out of vole and predator cycles? *Proceedings of the Royal Society of London (Biological Sciences)* 272 (1576):2045–2049.

Horton, S.P. 1996. Spotted Owls in managed forests of western Oregon and Washington. In *Raptors in Human Landscapes,* eds. D.M. Bird, D.E. Varland, and J.J. Negro, 215–231. San Diego, CA: Academic Press Limited.

Houston, C.S. 1966. Saskatchewan Long-eared Owl recovered in Mexico. *Blue Jay* 24:64.

Houston, C.S., and D.W. A. Whitfield. 1975. Eggs of other species in Great Horned Owl nests. *Auk* 92 (2):377–378.

Houston, C.S., D.G. Smith, and C. Rohner. 1998. Great Horned Owl *(Bubo virginianus).* In *The Birds of North America,* no. 372, eds. A. Poole and F. Gill. Philadelphia: Birds of North America, Inc.

Howell, S.N. G, and S. Webb. 1995. *A Guide to the Birds of Mexico and Northern Central America.* New York: Oxford University Press.

Hunt, W.G., D.E. Driscoll, E.W. Bianchi, and R.E. Jackman. 1992. *Ecology of Bald Eagles in Arizona, Part D: History of Nesting Population.* Report to U.S. Bureau of Reclamation, Contract 6-CS-30–04470. Santa Cruz, CA: BioSystems Analysis.

Hunter, J. E., D. Fix, G. A. Schmidt, and J. C. Power. 2005. *Atlas of the Breeding Birds of Humboldt County, California.* Eureka, CA: Redwood Region Audubon Society.

Hyde, P. S., and E. I. Knudsen. 2002. The optic tectum controls visually guided adaptive plasticity in the owl's auditory space map. *Nature* 415 (Jan. 3):73–76.

Ingles, L. G. 1965. *Mammals of the Pacific States.* Stanford, CA: Stanford University Press.

Jackson, R. W. 1925. Strange behavior of Great Horned Owl in behalf of young. *Auk* 42 (3):445.

James, P. C., G. A. Fox, and T. J. Ethier. 1990. Is the operational use of strychnine to control ground squirrels detrimental to burrowing owls? *Journal of Raptor Research* 24 (4):120–123.

Johnsgard, P. A. 2002. *North American Owls: Biology and Natural History.* Washington, DC: Smithsonian Institution Press.

Johnson, H. D. 2004. Owls were ignored in marsh restoration. *California Coast and Ocean* 20 (2). www.coastalconservancy.ca.gov/coast&ocean/summer2004/pages/view.htm.

Johnston, R. F. 1956. Predation by Short-eared Owls on a *Salicornia* salt marsh. *Wilson Bulletin* 68:91–102.

Karalus, K., and A. W. Eckert. 1974. *The Owls of North America.* New York: Doubleday and Co.

Kaufman, K. 2002. A little night magic. *Audubon Magazine,* January–February 1:74–75.

Keran, D. 1981. The incidence of man-caused and natural mortalities to raptors. *Journal of Raptor Research* 15:108–112.

Kerlinger, P., and M. R. Lein. 1988. Population ecology of Snowy Owls during winter on the Great Plains of North America. *Condor* 90: 866–874.

Kerlinger, P., M. R. Lein, and B. J. Sevick. 1985. Distribution and population fluctuations of wintering Snowy Owls *(Nyctea scandiaca)* in North America. *Canadian Journal of Zoology* 63:1829–1834.

Kidd, J. W., P. H. Bloom, C. W. Barrows, and C. T. Collins. In press. Status of Burrowing Owls in southwestern California. In *Proceedings of the California Burrowing Owl Symposium, 2003,* eds. J. H. Barclay, J. L. Lincer, J. Linthicum, K. Hunting, and T. A. Roberts. Point Reyes Station, CA: Institute for Bird Populations and Albion Environmental, Inc.

Klute, D. S., L. W. Ayers, M. T. Green, W. H. Howe, S. L. Jones, J. A. Shaffer, S. R. Sheffield, and T. S. Zimmerman. 2003. *Status Assessment and Conservation Plan for the Western Burrowing Owl in*

the United States. Biological Technical Publication FWS/BTP-R60001–2003. Washington DC: U.S. Department of the Interior, Fish and Wildlife Service.
Knudsen, E.I. 1981. The hearing of the barn owl. *Scientific American* 245:112–125.
Konishi, M. 1973. How the owl tracks its prey. *American Scientist* 61:414–424.
Konishi, M. 1983. Night owls are good listeners. *Natural History* 92 (9):56–59.
Konishi, M. 1993. Listening with two ears. *Scientific American* 268 (4):66–73.
Korbel, R. 1998. Erkrankungen des Augenhintergrundes bei Greifvögeln. In *Greifvögel und Falknerei: Jahrbuch des Deutschen Falkenordens 1997*, 69–88. Melsungen, Germany: Verlag J. Neumann-Neudamm GmbH and Co.
LaHaye, W.S., R.J. Gutiérrez, and D.R. Call. 1997. Nest-site selection and reproductive success of California Spotted Owls. *Wilson Bulletin* 109 (1):42–51.
Lane, P.A., and J.R. Duncan. 1987. Observations of Northern Hawk-Owls nesting in Roseau County. *Loon* 59:165–174.
Larsen, C.J. 1987. *Short-eared Owl Breeding Survey.* Project IW-65-R-4, Nongame Wildlife Investigations, Job II-13, Final Report. Sacramento: California Department of Fish and Game.
Latta, F.F. 1949. *Handbook of Yokuts Indians.* Oildale, CA: Bear State Books.
Laymon, S.A. 1989. Altitudinal migration movements of Spotted Owls in the Sierra Nevada, California. *Condor* 91 (4):837–841.
Levey, D.J., R.S. Duncan, and C.F. Levins. 2004. Use of dung as a tool by Burrowing Owls. *Nature* 431 (Sept. 2):39.
Ligon, J.D. 1968. *The biology of the Elf Owl* (Micrathene whitneyi). Miscellaneous Publication 136. Ann Arbor: University of Michigan Museum of Zoology.
Lilley, G.M. 1998. *A Study of the Silent Flight of the Owl.* Reston, VA: American Institute of Aeronautics and Astronautics.
Linkenhoker, B.A., and E.I. Knudsen. 2002. Incremental training increases the plasticity of the auditory space map in adult Barn Owls. *Nature* 419 (Sept. 19):293–296.
Linkhart, B.D., and R.T. Reynolds. 1997. Territories of Flammulated Owls *(Otus flammeolus)*: Is occupancy a measure of habitat quality? In *Biology and Conservation of Owls in the Northern Hemisphere: 2nd International Symposium* (February 5–9, Winnipeg, MB),

General Technical Report NC190, eds. J. R. Duncan, D. H. Johnson, and T.H. Nicholls, 250–254. St. Paul, MN: U.S. Department of Agriculture Forest Service, North Central Research Station.

Linkhart, B.D., and R.T. Reynolds. 2006. Lifetime reproduction of Flammulated Owls in Colorado. *Journal of Raptor Research* 40 (1):29–37.

Lockie, J.D. 1955. The breeding habits and food of Short-eared Owls after a vole plague. *Bird Study* 2:53–67.

Loyd, L.R.W. 1927. *Bird Facts and Fallacies.* London: Hutchinson.

Marcot, B.G., and R. Hill. 1980. Flammulated Owls in northwestern California. *Western Birds* 11:141–149.

Margolin, M. 1978. *The Ohlone Way.* Berkeley, CA: Heyday Books.

Margolin, M., ed. 1993. *The Way We Lived.* Berkeley, CA: Heyday Books.

Marks, J.S. 1986. Nest-site characteristics and reproductive success of Long-eared Owls in southwestern Idaho. *Wilson Bulletin* 98 (4):547–560.

Marks, J.S. 1997. Is the Northern Saw-whet Owl *(Aegolius acadicus)* nomadic? In *Biology and Conservation of Owls in the Northern Hemisphere: 2nd International Symposium* (February 5–9, Winnipeg, MB), General Technical Report NC190, eds. J.R. Duncan, D.H. Johnson, and T.H. Nicholls, 260. St. Paul, MN: U.S. Department of Agriculture Forest Service, North Central Research Station.

Marks, J.S., and J.H. Doremus. 2000. Are Northern Saw-whet Owls nomadic? *Journal of Raptor Research* 34 (4):299–304.

Marks, J.S., and A.E. H. Perkins. 1999. Double brooding in the Long-eared Owl. *Wilson Bulletin* 111 (2):273–276.

Marks, J.S., D.L. Evans, and D.W. Holt. 1994. Long-eared Owl *(Asio otus).* In *The Birds of North America,* no. 133, eds. A. Poole and F. Gill. Philadelphia: The Academy of Natural Sciences and The American Ornithologists' Union.

Marks, J.S., R.J. Cannings, and H. Mikkola. 1999. Family Strigidae (typical owls). In *Handbook of the Birds of the World,* vol. 5, eds. J. del Hoyo, A. Elliott, and J. Sargatal, 76–151. Barcelona: Lynx Edicions.

Marks, J.S., J.L. Dickinson, and J. Haydock. 1999. Genetic monogamy in Long-eared Owls. *Condor* 101 (4):854–859.

Marshall, J.T. 1939. Territorial behavior of the Flammulated Screech Owl. *Condor* 41 (2):71–78.

Martell, M.S., J. Schladweiler, and F. Cuthbert. 2001. Status and attempted reintroduction of Burrowing Owls in Minnesota, U.S.A. *Journal of Raptor Research* 35 (4):331–336.

Marti, C.D. 1973. Food consumption and pellet formation rates in four owl species. *Wilson Bulletin* 85 (2):178–181.

Marti, C.D. 1974. Feeding ecology of four sympatric owls. *Condor* 76:45–61.

Marti, C.D. 1976. A review of prey selection by the Long-eared Owl. *Condor* 78:331–336.

Marti, C.D. 1987. Raptor food habits studies. In *Raptor Management Techniques Manual,* eds. B.A. Giron-Pendleton, B.A. Millsap, K.W. Cline, and D.M. Bird, 67–80. Washington, DC: National Wildlife Federation.

Marti, C.D. 1990. Same-nest polygyny in the Barn Owl. *Condor* 92 (1):261–263.

Marti, C.D. 1992. Barn Owl *(Tyto alba).* In *The Birds of North America,* no. 1, eds. A. Poole, P. Stettenheim, and F. Gill. Philadelphia: The Academy of Natural Sciences and The American Ornithologists' Union.

Marti, C.D. 1997a. Lifetime reproductive success in Barn Owls near the limit of the species' range. *Auk* 114 (4):581–592.

Marti, C.D. 1997b. Flammulated Owls *(Otus flammeolus)* breeding in deciduous forests. In *Biology and Conservation of Owls in the Northern Hemisphere: 2nd International Symposium* (February 5–9, Winnipeg, MB), General Technical Report NC190, eds. J.R. Duncan, D.H. Johnson, and T.H. Nicholls, 262–266. St. Paul, MN: U.S. Department of Agriculture Forest Service, North Central Research Station.

Marti, C.D. 1999. Natal and breeding dispersal in Barn Owls. *Journal of Raptor Research* 33 (3):181–189.

Marti, C.D., and J.G. Hogue. 1979. Selection of prey by size in screech owls. *Auk* 96 (2):319–327.

Marti, C.D., and M.N. Kochert. 1995. Are Red-tailed Hawks and Great Horned Owls diurnal-nocturnal dietary counterparts? *Wilson Bulletin* 107 (4):615–628.

Marti, C.D., and P.W. Wagner. 1985. Winter mortality in Common Barn-Owls and its effect on population density and reproduction. *Condor* 87 (1):111–115.

Martin, D.J. 1971. Unique Burrowing Owl pellets. *Journal of Field Ornithology* 42 (4):298–299.

Martin, S.J. 1983. Burrowing Owl occurrence on White-tailed Prairie Dog colonies. *Journal of Field Ornithology* 54:422–423.

Massemin, S., and Y. Handrich. 1997. Higher winter mortality of the Barn Owl compared to the Long-eared Owl and the Tawny Owl: Influence of lipid reserves and insulation? *Condor* 99 (4):969–971.

Mazur, K.M., and P.C. James. 2000. Barred Owl *(Strix varia).* In *The Birds of North America,* no. 508, eds. A. Poole and F. Gill. Philadelphia: Birds of North America, Inc.

McCallum, D.A. 1994a. Flammulated Owl *(Otus flammeolus).* In *The Birds of North America,* no. 93, eds. A. Poole and F. Gill. Philadelphia: The Academy of Natural Sciences and The American Ornithologists' Union.

McCallum, D.A. 1994b. Review of technical knowledge: Flammulated Owls. In *Flammulated, Boreal, and Great Gray Owls in the United States: A Technical Conservation Assessment,* General Technical Report GTR-RM-253, eds. G.D. Hayward and J. Verner, 14–46. Fort Collins, CO: U.S. Department of Agriculture Forest Service.

McCallum, D.A., F.R. Gehlbach, and S.W. Webb. 1995. Life history and ecology of Flammulated Owls in a marginal New Mexico population. *Wilson Bulletin* 107 (3):530–537.

McGarigal, K., and J.D. Fraser. 1985. Barred Owl responses to recorded vocalizations. *Condor* 87:552–553.

Mehlman, D.W. 1997. Change in avian abundance across the geographic range in response to environmental change. *Ecological Applications* 7 (2):614–624.

Mikkola, H. 1983. *Owls of Europe.* Vermillion, SD: Buteo Books.

Millsap, B.A. 1998. Long-eared Owl. In *Raptors of Arizona,* ed. R.L. Glinski, 175–177. Tucson: University of Arizona Press.

Millsap, B.A., and K.N. Woodruff. 1979. Burrow nesting by Barn Owls in northcentral Colorado. Paper presented at the 1979 Raptor Research Annual Meeting, Davis, CA.

Mineau, P., M.R. Fletcher, L.C. Glaser, N.J. Thomas, C. Brassard, L.K. Wilson, J.E. Elliott, L.A. Lyon, C.J. Henny, T. Bollinger, and S.L. Porter. 1999. Poisoning of raptors with organophosphorus and carbamate pesticides with emphasis on Canada, U.S., and U.K. *Journal of Raptor Research* 33 (1):1–37.

Montevecchi, W.A., and A.D. Maccarone. 1987. Differential effects of a Great Horned Owl decoy on the behavior of juvenile and adult Gray Jays. *Journal of Field Ornithology* 58 (2):148–151.

Moser, J.A., and C.J. Henry. 1976. Thermal adaptiveness of plumage color in screech owls. *Auk* 93:614–619.

Mueller, H.C. 1986. The evolution of reversed sexual dimorphism in owls: An empirical analysis of possible selective factors. *Wilson Bulletin* 98:387–406.

Mueller, H.C., and D.D. Berger. 1959. Some long distance Barn Owl recoveries. *Journal of Field Ornithology* 30 (3):182.

Murphy, R.K. 1992. Long-eared Owl ingests nestlings' feces. *Wilson Bulletin* 104 (1):192–193.

Murray, J.O. 2005. The influence of grazing treatments on density of nesting Burrowing Owls on the Cheyenne River Sioux Reservation. M.S. thesis. Brookings: South Dakota State University. Abstract in *Wingspan* 15 (1):13–14.

Nero, R.W. 1980. *The Great Gray Owl: Phantom of the Northern Forest.* Washington, DC: Smithsonian Institution Press.

Nero, R.W. 1995. Notes on a wintering Northern Hawk Owl in Manitoba. *Blue Jay* 53:205–214.

Nicholls, T.H., and D.W. Warner. 1972. Barred Owl habitat use as determined by radiotelemetry. *Journal of Wildlife Management* 36:213–224.

Noble, P.L. 1990. Distribution and density of owls at Monte Bello Open Space Preserve, Santa Clara County, California. *Western Birds* 21:11–16.

Nuijen, H. 1992. Ransuilen *Asio otus* nemen Zandbad. *Limosa* 65:125.

Oleyar, M.D., C.D. Marti, and M. Mika. 2003. Vertebrate prey in the diet of Flammulated Owls in northern Utah. *Journal of Raptor Research* 37 (3):244–246.

Otteni, L.C., E.G. Bolen, and C. Cottam. 1972. Predator-prey relationships and reproduction of the Barn Owl in southern Texas. *Wilson Bulletin* 84 (4):434–448.

Page, G.W., and D.F. Whitacre. 1975. Raptor predation on wintering shorebirds. *Condor* 77 (1):73–83.

Palmer, D.A. 1986. Habitat selection, movements, and activity of Boreal and Saw-whet Owls. M.S. thesis. Fort Collins: Colorado State University.

Parmelee, D.F. 1972. Canada's incredible arctic owls. *Beaver,* Summer, 30–41.

Parmelee, D.F. 1992. Snowy Owl *(Nyctea scandiaca).* In *The Birds of North America,* no. 10, eds. A. Poole, P. Stettenheim, and F. Gill. Philadelphia: The Academy of Natural Sciences and The American Ornithologists' Union.

Payne, R.S. 1962. How the Barn Owl locates prey by hearing. *Living Bird* 1:151–159.

Peeters, H.J. 1963. Two observations of avian predation. *Wilson Bulletin* 75 (3):274.

Peeters, H.J. 1994. Suspected poisoning of Golden Eagles *Aquila chrysaetos* by chlorophacinone. In *Raptor Conservation Today,* eds. B.-U. Meyburg and R. D. Chancellor, 775–776. Berlin: World Working Group on Birds of Prey, The Pica Press.

Peeters, H., and P. Peeters. 2005. *Raptors of California.* Berkeley and Los Angeles: University of California Press.

Penteriani, V., C. Alonso-Alvarez, M. del Mar Delgado, F. Sergio, and M. Ferrer. 2006. Brightness variability in the white badge of the Eagle Owl *(Bubo bubo). Journal of Avian Biology* 37 (1):110–116.

Pezzolesi, L.S.W., and R.S. Lutz. 1994. Foraging and crepuscular/nocturnal behaviors of the Western Burrowing Owl. Abstract. *Journal of Raptor Research* 28 (1):63–64.

Phelan, F.J.S. 1977. Food caching in the screech owl. *Condor* 79 (1): 127.

Philips, J.R. 2000. A review and checklist of the parasitic mites (Acarina) of the Falconiformes and Strigiformes. *Journal of Raptor Research* 34 (3):210–231.

Plumpton, D.L., and R.S. Lutz. 1994. Sexual size dimorphism, mate choice, and productivity of Burrowing Owls. *Auk* 111 (3):724–727.

Point Reyes Bird Observatory. Safe nest boxes for owls in the West. www.prbo.org/cms/docs/edu/Owl_nest_boxes.pdf.

Powell, J.A., and C.L. Hogue. 1979. *California Insects.* Berkeley and Los Angeles: University of California Press.

Proudfoot, G.A., and S.L. Beasom. 1996. Responsiveness of Cactus Ferruginous Pygmy-Owls to broadcasted conspecific calls. *Wildlife Society Bulletin* 24:294–297.

Proudfoot, G.A., and S.L. Beasom. 1997. Food habits of nestling Ferruginous Pygmy-Owls in southern Texas. *Wilson Bulletin* 109 (4):741–748.

Proudfoot, G.A., and R.R. Johnson. 2000. Ferruginous Pygmy-Owl *(Glaucidium brasilianum).* In *The Birds of North America,* no. 498, eds. A. Poole and F. Gill. Philadelphia: Birds of North America, Inc.

Rains, C. 1997. Comparison of food habits of the Northern Saw-whet Owl *(Aegolius acadicus)* and the Western Screech-Owl *(Otus kennicottii)* in southwestern Idaho. In *Biology and Conservation of Owls in the Northern Hemisphere: 2nd International Symposium* (February 5–9, Winnipeg, MB), General Technical Report NC190, eds. J. R. Duncan, D. H. Johnson, and T. H. Nicholls, 339–346. St. Paul, MN: U.S. Department of Agriculture Forest Service, North Central Research Station.

Reynolds, R. T., and B. D. Linkhart. 1987. Fidelity to territory and mate in Flammulated Owls. In *Biology and Conservation of Northern Forest Owls*, General Technical Report RM-142, eds. R. W. Nero, R. J. Clark, R. J. Knapton, and R. H. Hamre, 234–238. Washington, DC: U.S. Department of Agriculture Forest Service.

Reynolds, R. T., and B. D. Linkhart. 1998. Flammulated Owl. In *Raptors of Arizona*, ed. R. Glinski. Tucson: University of Arizona Press.

Richmond, M. L., L. R. DeWeese, and R. E. Pillmore. 1980. Brief observations on the breeding biology of the Flammulated Owl in Colorado. *Western Birds* 11:35–46.

Roberson, D. 2002. *Monterey Birds*. Pacific Grove, CA: Monterey Peninsula Audubon Society.

Robinson, M., and C. D. Becker. 1986. Snowy Owls on Fetlar. *British Birds* 78:228–242.

Robinson, T. S. 1954. Cannibalism by a Burrowing Owl. *Wilson Bulletin* 66 (1):72.

Rohner, C. 1995. Responses of Great Horned Owls *(Bubo virginianus)* to the Snowshoe Hare cycle in the boreal forest. Abstract. *Journal of Raptor Research* 29 (1):65.

Rohner, C., J. N. M. Smith, J. Stroman, M. Joyce, F. I. Doyle, and R. Boonstra. 1995. Northern Hawk-Owls in the Nearctic boreal forest: Prey selection and population consequences of multiple prey cycles. *Condor* 97 (1):208–220.

Rohner, C., C. J. Krebs, D. B. Hunter, and D. C. Currie. 2000. Roost site selection of Great Horned Owls in relation to black fly activity: An anti-parasite behavior? *Condor* 102 (4):950–955.

Rosenberg, D. K., and K. L. Haley. 2004. The ecology of Burrowing Owls in the agroecosystem of the Imperial Valley, California. *Studies in Avian Biology* 27:120–135.

Rosier, J. R., N. A. Ronan, and D. K. Rosenberg. 2006. Post-breeding dispersal of Burrowing Owls in an extensive California grassland. *American Midland Naturalist* 155 (1):162–167.

Rucci, M., G. M. Edelman, and J. Wray. 1999. Adaptation of orienting behavior: From the Barn Owl to a robotic system. *IEEE Transactions on Robotics and Automation* 15 (1):96–111.

Rudolph, S. G. 1978. Predation ecology of coexisting Great Horned and Barn owls. *Wilson Bulletin* 90 (1):134–137.

San Francisco Estuary Institute. 1999. *Baylands Ecosystem Habitat Goals Report.* Oakland, CA: San Francisco Bay Regional Water Quality Control Board. Also available at www.sfei.org.

Sater, D.M., E.D. Forsman, F.L. Ramsey, and E.M. Glenn. 2006. Distribution and habitat associations of Northern Pygmy-Owls in Oregon. *Journal of Raptor Research* 40 (2):89–97.

Sauer, J.R., J.E. Hines, and J. Fallon. 2005. *The North American Breeding Bird Survey, Results and Analysis 1966–2004.* Version 2005.2. Laurel, MD: U.S. Geological Survey Patuxent Wildlife Research Center.

Schlitter, D.A. 1973. A new species of gerbil from south west Africa with remarks on *Gerbillus tytonis* Bauer and Neithamer, 1959 (Rodentia: Gerbillinae). *Bulletin of the Southern California Academy of Sciences* 72 (1):13–18.

Schmutz, J.K. G. Wood, and D. Wood. 1991. Spring and summer prey of Burrowing Owls in Alberta. *Blue Jay* 49:93–97.

Schueler, F.W. 1972. A new method of preparing owl pellets: Boiling in NaOH. *Journal of Field Ornithology* 43 (2):142.

Sheffield, S.R. 1997. Current status, distribution, and conservation of the Burrowing Owl *(Speotyto cunicularia)* in midwestern and western North America. In *Biology and Conservation of Owls of the Northern Hemisphere: 2nd International Symposium,* (February 5–9, Winnipeg, MB), eds. J.R. Duncan, D.H. Johnson, and T.H. Nicholls, 399–407. General Technical Report NC-190. St. Paul, MN: U.S. Department of Agriculture Forest Service, North Central Forest Experiment Station.

Sleep, D.J. H., and R.D. H. Barrett. 2004. Insect hawking observed in the Long-eared Owl *(Asio otus). Journal of Raptor Research* 38 (4): 379–380.

Small, A. 1994. *California Birds: Their Status and Distribution.* Vista, CA: Ibis Publishing.

Smith, B.W., and J.R. Belthoff. 2001a. Identification of ectoparasites on Burrowing Owls in southwestern Idaho. *Journal of Raptor Research* 35 (2):159–161.

Smith, B.W., and J.R. Belthoff. 2001b. Burrowing Owls and development: Short-distance nest burrow relocation to minimize construction impacts. *Journal of Raptor Research* 35 (4):385–391.

Smith, D.G., and C.D. Marti. 1976. Distributional status and ecology of Barn Owls in Utah. *Journal of Raptor Research* 10:33–44.

Smith, D.G., and B.A. Smith. 1972. Hunting methods and success of newly fledged Great Horned Owls. *Journal of Field Ornithology* 43 (2):142.

Smith, D.G., A. Devine, and D. Gendron. 1982. An observation of copulation and allopreening of a pair of Whiskered Owls. *Journal of Field Ornithology* 53:51–52.

Smith, D.G., A. Devine, and D. Devine. 1983. Observations of fishing by a Barred Owl. *Journal of Field Ornithology* 54:88–89.

Smith, M.D., and C.J. Conway. 2007. Use of mammal manure by nesting Burrowing Owls: A test of four functional hypotheses. *Animal Behaviour* 73 (1):65–73.

Solis, D.M., and R.J. Gutierrez. 1990. Summer habitat ecology of Northern Spotted Owls in northwestern California. *Condor* 92 (3):739–748.

Sovern, S.G., E.D. Forsman, B.L. Biswell, D.N. Rolph, and M. Taylor. 1994. Diurnal behavior of the Spotted Owl in Washington. *Condor* 96:200–202.

Sproat, T.M., and G. Ritchison. 1994. The antipredator vocalizations of adult Eastern Screech-Owls. *Journal of Raptor Research* 28 (2): 93–99.

Sprunt, A. 1943. Generally unrecognized habit of Florida Burrowing Owls. *Auk* 60 (1):97–98.

Sprunt, A. 1955. *North American Birds of Prey.* New York: Harper and Brothers.

Stahlecker, D.W., and R.B. Duncan. 1996. The Boreal Owl at the southern terminus of the Rocky Mountains: Undocumented long-time resident or recent arrival? *Condor* 98 (1):153–161.

Steinhart, P. 1990. *California's Wild Heritage: Threatened and Endangered Animals in the Golden State.* Sacramento: California Department of Fish and Game; San Francisco: California Academy of Sciences and Sierra Club Books.

Stock, S.L., P.J. Heglund, G.S. Kaltenecker, J.D. Carlisle, and L. Leppert. 2006. Comparative ecology of the Flammulated Owl and Northern Saw-whet Owl during fall migration. *Journal of Raptor Research* 40 (2):120–129.

Sutton, G.M. 1929. Insect-catching tactics of the Screech Owl. *Auk* 46 (4):545–546.

Sutton, G.M. 1932. The birds of Southhampton Island. *Memoirs of the Carnegie Museum* 12 (part 2, sect. 2):1–275.

Swarthout, E.C.H., and R.J. Steidl. 2003. Experimental effects of hiking on breeding Mexican Spotted Owls. *Conservation Biology* 17 (1):307–315.

Templeton, C.N., E. Greene, and K. Davis. 2005. Allometry of alarm calls: Black-capped Chickadees encode information about predator size. *Science* 308 (June 24):1934–1937.

Terrill, S.B., and P. Delevoryas. 1993. Relocation of Burrowing Owls during courtship period. Abstract. *Journal of Raptor Research* 27 (1):93–94.

Thomas, R.H., M.M. Choudhari, and R.D. Joslin. 2002. *Flow and Noise Control: Review and Assessment of Future Directions.* NASA-TM-2002–211631. Hampton, VA: National Aeronautics and Space Administration, Langley Research Station.

Thompson, E.E. 1891. *The Birds of Manitoba.* U.S. National Museum no. 841. Washington, DC: Smithsonian Institution.

Thompson, L. 1991. *To the American Indian: Reminiscences of a Yurok Woman.* Berkeley, CA: Heyday Books.

Thomsen, L. 1971. Behavior and ecology of Burrowing Owls on the Oakland municipal airport. *Condor* 73:177–192.

Trulio, L.A. 1995. Passive relocation: A method to preserve Burrowing Owls on disturbed sites. *Journal of Field Ornithology* 66 (1):99–106.

Trulio, L.A. 1997. Strategies for protecting Western Burrowing Owls *(Speotyto cunicularia hypugaea)* from human activities. In *Biology and Conservation of Owls in the Northern Hemisphere: 2nd International Symposium* (February 5–9, Winnipeg, MB), General Technical Report NC190, eds. J.R. Duncan, D.H. Johnson, and T.H. Nicholls, 461–465. St. Paul, MN: U.S. Department of Agriculture Forest Service, North Central Research Station.

Trulio, L.A. 2004. A message from the Burrowing Owl. *California Coast and Ocean* 20 (2). www.coastalconservancy.ca.gov/coast&ocean/summer2004/pages/view.htm.

Turner, J.C., Jr., and L. McClanahan Jr. 1981. Physiogenesis of endothermy and its relation to growth in the Great Horned Owl, *Bubo virginianus. Comparative Biochemistry and Physiology* 68A: 167–173.

U.S. Fish and Wildlife Service. 2005. *Seabird Conservation Plan—Pacific Region.* www.fws.gov/pacific/migratorybirds/Seabird_Conservation_Plan_Webpages/Complete_USFWS_Seabird_Conservation_Plan.pdf.

Verner, J. 1992. Data needs for avian conservation biology: Have we avoided critical research? *Condor* 94 (1):301–303.

Verner, J., and A.S. Boss, tech. coordinators. 1980. *California Wildlife and Their Habitats: Western Sierra Nevada.* General Technical Report PSW-37. Berkeley, CA: U.S. Department of Agriculture Forest Service, Pacific Southwest Forest and Range Experimental Station.

Verner, J., K.S. McKelvey, B.R. Noon, R.J. Gutiérrez, G.I. Gould, and T.W. Beck, tech. coordinators. 1992. *The California Spotted Owl: A Technical Assessment of its Current Status.* General Technical Report PSW-GTR-133. Albany, CA: U.S. Department of Agriculture Forest Service, Pacific Southwest Research Station.

Voous, K.H. 1989. *Owls of the Northern Hemisphere.* Cambridge, MA: MIT Press.

Walk, J.W., T.L. Esker, and S.A. Simpson. 1999. Continuous nesting of Barn Owls in Illinois. *Wilson Bulletin* 111 (4):572–573.

Ward, J.P., R.J. Gutiérrez, and B.R. Noon. 1998. Habitat selection by Northern Spotted Owls: The consequences of prey selection and distribution. *Condor* 100 (1):79–92.

Watson, A. 1957. The behaviour, breeding, and food-ecology of the Snowy Owl *(Nyctea scandiaca). Ibis* 99:419–462.

Whalen, D.M., B.D. Watts, and D.W. Johnston. 2000. Diet of autumn migrating Northern Saw-whet Owls on the eastern shore of Virginia. *Journal of Raptor Research* 34 (1):42–44.

White, K. 1996. Comparison of fledging success and sizes of prey consumed by Spotted Owls in northwestern California. *Journal of Raptor Research* 30 (4):234–236.

Wink, M., and P. Heidrich. 1999. Molecular evolution and systematics of the owls (Strigiformes). In *Owls: A Guide to the Owls of the World,* eds. C. König, F. Weick, and J.H. Becking, 39–57. New Haven, CT: Yale University Press.

Winter, J. 1985. *Great Gray Owl Survey, 1984.* Project W-65-R-2 (554), Wildlife Management Branch, Nongame Wildlife Investigations, Job II-3, Progress Report. Sacramento: California Department of Fish and Game.

Yaffee, S.L. 1994. *The Wisdom of the Spotted Owl: Policy Lessons for a New Century.* Washington, DC: Island Press.

Zarn, M. 1974. *Burrowing Owl.* Report no. 11. Habitat Management Series for Unique or Endangered Species. Denver, CO: Bureau of Land Management.

Zeiner, D.C., W.F. Laudenslayer Jr., K.E. Mayer, and M. White, eds. 1990. *California's Wildlife.* Vol. 2, *Birds.* Sacramento, CA: California Department of Fish and Game Resources Agency.

INDEX

Page numbers in **boldface type** indicate main discussions of owl species.

ADDITIONAL CAPTIONS

PAGE X The thickly layered plumage of a Great Horned Owl gives little clue to the actual size of its body.

PAGES XVII–1 Compared to the body feathers of other birds, those of owls have extraordinary long, downy bases (Great Horned Owl feather shown).

PAGES 10–11 This Northern Pygmy-Owl's posture indicates that it is hunting (in broad daylight).

PAGES 52–53 On the cusp of prefledging these young Northern Saw-whets have been removed temporarily from their nest box (originally intended for Wood Ducks) for banding.

PAGE 274 Northern Pygmy-Owls, being chiefly sight hunters, have poorly developed facial disks.

PAGE 280 A Great Gray Owl blends in well with a burnt-over lodgepole pine forest.

PAGE 304 Unlike many hawks, the great majority of owls prefers to use concealed perches, as does this Great Horned Owl.

ABOUT THE AUTHOR

Photo by Julian Peeters

Hans J. Peeters is an ornithologist whose main interest is raptors. Born in Germany in 1937, he came to the United States at the age of 16, received a B.A. in Comparative Literature at the University of California at Berkeley, and went on to complete a graduate degree in zoology at the same institution. For 37 years he taught subjects ranging from ecology to zoology and field biology at Chabot College in Hayward, California. He is coauthor of two other University of California Press natural history guides, *Raptors of California* (with his wife, Pam Peeters) and *Mammals of California* (with E.W. Jameson Jr., with whom he is also coauthor of two books on falconry).

Over the years, his avocation of painting developed into a second career, and he much enjoys combining his interests in science and art wherever possible. He has painted endangered species for conservation postage stamps for Mexico, and other conservation work includes painting a poster of a Harpy Eagle for the country of Panama, where this national bird is endangered.

Hans and Pam live in Sunol, California. Their acre and a half of oak-bay and riparian woodland property is home to and visited by several species of owls, but it was not until Hans signed a contract to write this book that a Northern Pygmy-Owl, whose habits are not well known, first appeared in the yard. The little owl then proceeded to find a mate and raise a family!

Hans has watched owls nearly all over the world. His interest in these birds came early: as a 10-year-old, he hung above his bed a Tawny Owl he had stuffed—incorrectly, it turned out, for maggots kept appearing in the sheets. In this photo in Malaysia, an Oriental Bay Owl and a Collared Scops-Owl use him as a perch, while a Barred Eagle-Owl allopreens his ear.

Series Design: Barbara Jellow
Design Enhancements: Beth Hansen
Design Development: Jane Tenenbaum
Cartographer: Dartmouth Publishing
Composition: Jane Tenenbaum
Indexer: Jean Mann
Text: 9/10.5 Minion
Display: Franklin Gothic Book and Demi
Printer and binder: Golden Cup Printing Company Limited